Home Security Projects

The Maplin series

This book is part of an exciting series developed by Butterworth-Heinemann and Maplin Electronics Plc. Books in the series are practical guides which offer electronic constructors and students clear introductions to key topics. Each book is written and compiled by a leading electronics author.

Other books published in the Maplin series include:

Computer Interfacing	Graham Dixey	0 7506 2123 0
Logic Design	Mike Wharton	0 7506 2122 2
Music Projects	R A Penfold	0 7506 2119 2
Starting Electronics	Keith Brindley	0 7506 2053 6
Audio IC Projects	Maplin	0 7506 2121 4
Auto Electronics Projects	Maplin	0 7506 2296 2
Video and TV Projects	Maplin	0 7506 2297 0
Integrated Circuit Projects	Maplin	0 7506 2578 3
Test Gear and Measurements	Danny Stewart	0 7605 2601 1
Power Supply Projects	Maplin	0 7506 2602 X
The Maplin Approach to Professional Audio	T.A. Wilkinson	0 7506 2120 6

Home Security Projects

Newnes
An imprint of Butterworth-Heinemann Ltd
Linacre House, Jordan Hill, Oxford OX2 8DP

Ⓡ A member of the Reed Elsevier group

OXFORD   LONDON   BOSTON
MUNICH   NEW DELHI   SINGAPORE   SYDNEY
TOKYO   TORONTO   WELLINGTON

British Library Cataloguing in Publication Data
A catalogue record for this book is available from the
British Library
ISBN 0 7506 2603 8

Library of Congress Cataloguing in Publication Data
A catalogue record for this book is available from the
Library of Congress

 Edited by Co-publications, Loughborough

 Typeset and produced by Sylvester North, Sunderland

all part of The Sylvester Press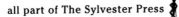

Printed in Great Britain

Contents

Preface

This book is a collection of projects previously published in *Electronics — The Maplin Magazine.*

Each project is selected for publication because of its special features, because it is unusual, because it electronically clever, or simply because we think readers will be interested in it. Some of the devices used are fairly specific in function — in other words, an integrated circuit used is designed and built for one purpose alone. Others, on the other hand, are not specific at all, and can be used in any number of applications.

This is just one of the Maplin series of books published by Newnes books covering all aspects of computing and electronics. Others in the series are available from all good bookshops.

Maplin Electronics Plc supplies a wide range of electronics components and other products to private individuals and trade customers. Telephone: (01702) 552911 or write to Maplin Electronics, PO Box 3, Rayleigh, Essex SS6 8LR, for further details of product catalogue and locations of regional stores.

1 Alarms

Explosive gas alarm

Dangerous gas leaks, particularly in confined spaces, causing explosions and fires, are becoming a more common occurrence, usually damaging property and often maiming or even killing people. This gas detector has been designed to prevent the build-up of these gases by sounding a loud alarm before sufficient gas has leaked to cause a damaging explosion. The sensor used consists of two separate units, the sensor itself and a reference compensator. Both elements are connected in series and used to form two legs of a Wheatstone bridge. The two elements have similar resistance under normal conditions and vary equally with changes of ambient temperature, maintaining the bridge in balance. The presence of an inflammable gas causes the sensor element to increase in temperature, due to the oxidisation of the gas on the surface of its platinum heating element. This increase in temperature causes an increase in resistance of the element and thus the bridge becomes unbalanced, the detection of which causes the alarm to sound.

The fairly high current (about 400 mA) required by the sensors and their associated circuitry make it undesirable to have the sensors permanently energised, particularly when installed in a boat or caravan where the power is supplied from a 12 V battery. This problem is overcome by testing for gas once every 5 or 6 minutes and latching the alarm on when gas is detected. This test period is adequate because in most cases the build up of gas due to a leak is fairly slow and the alarm should operate well before a dangerous level is reached.

The system will detect all common explosive or inflammable gases such as butane, propane, methane, town gas, natural gas, and petrol vapour. The sensors are enclosed in double wire mesh housings to prevent any chance of the sensor itself igniting any gases encountered.

How it works

The Wheatstone bridge, previously mentioned, consists of R2, R4, RV1, and the two sensing elements. The balancing of the bridge is performed by adjusting RV1. The CAL switch (S1) is used to unbalance the bridge by a small amount, simulating the presence of a small amount of gas, this being used in setting up the alarm for maximum sensitivity. The state of the bridge is monitored by the dual op-amp, IC1a and IC1b, whose output is used to turn on TR3 under alarm conditions. IC4 forms a dual oscillator to produce the warbling alarm tone, and its output is fed to TR4 which drives the piezo tweeter which provides the audio output. The alarm, when activated, latches on and sounds continuously until reset by the action of S3, which also disconnects the speaker for testing purposes. Latching is performed by connecting the positive voltage fed from the collector of TR3 (under alarm conditions) back to the inverting input of IC1a via D3, D4 and R5.

The sequence timing of the alarm is carried out by IC3, which is a 14-stage ripple counter with built in oscillator. The frequency of the oscillator is determined by R17, C3 and C4 running at about 1 cycle every 4 seconds. The

Home security projects

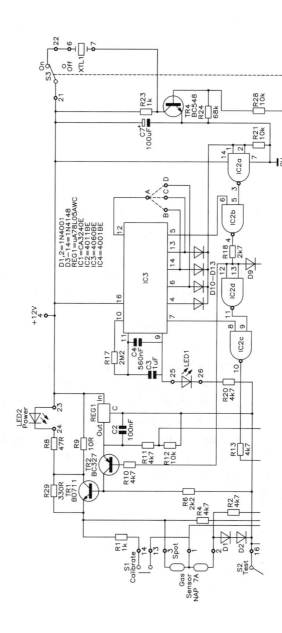

Figure 1.1 Circuit diagram of Gas Alarm

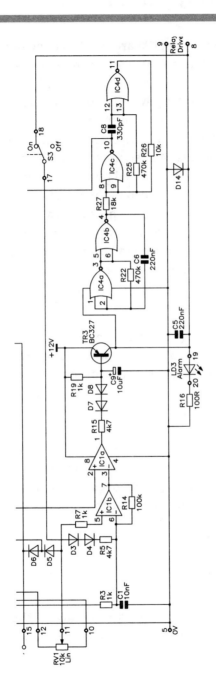

Figure 1.1 Continued

various outputs from IC3 are used to control the switching regulator TR1 and to enable the monitor circuit via IC2, R13 and D5. The sensors require at least 20 seconds to settle down after power is applied before a test can be made and for this reason the sensors are powered for about 80 seconds per test cycle, but the alarm is only enabled for the last 40 seconds of this period. The time between tests can be altered by selecting various straps but under normal conditions the shortest period is recommended. (Link A to B). The sensors require a stable supply independent of variations of the incoming supply voltage, this is achieved by the 5 V regulator (REG1) which provides the base current for the power switching transistor (TR1) via TR2, which forms part of the control circuit, fed from IC3.

The *test* switch (S2) overrides the timer connecting power to the sensors and also enabling the alarm circuit. LED2 lights whenever current is drawn by the sensors and will be on continuously when S2 is in the *test* position. Note that when the alarm is working, LED2 will only be on for the 80 seconds of the test period during each test cycle. LED3 gives an indication of an alarm even when the *alarm cancel* switch (S3) is operated and is used for setting up purposes. LED1 flashes at clock rate and is an indication that the timer is running.

Construction

Construct the circuit board referring to the Parts List and component overlay on the board. Ensure the correct polarity of all diodes, transistors, integrated circuits and electrolytic capacitors. The sensors should be mounted carefully on their board, avoiding excessive heat and making sure that the one marked with the spot is in the correct position. The required amount of cable should be connected to the sensor board at this stage but not terminated on the main board before testing.

Setting up and testing

Connect the 10 ohm test resistor provided with the kit in place of the sensors (between pins 2 and 3). Switch S3 to *alarm off* and S2 to *test*. Connect 12–24 V to the unit and observe LED1; this should flash regularly at about once every 4 seconds. Connect a multimeter; set to a range that reads up to 10 V, between –V battery supply (pin 5) and pin 3 (also connected to one end of test resistor). A reading of between 4.1 and 4.6 volts should be obtained at this point. Transfer the meter to pin 2; a reading of 1.4 to 1.8 volts should be measured.

Note, if these readings are not correct *do not* connect the sensors. *Warning*, the test resistor will become *hot* during this test.

When the above conditions are correct, disconnect the supply and remove the test resistor. Connect the sensor cable ensuring correct location of wires. Reapply power and check that LED2 (green) is on. When current is first applied to the sensors, a slight smell of burning may be noticed; this is quite normal. Turn RV1 fully clockwise; in this position LED3 (red) should be out. Wait 20 seconds then rotate RV1 slowly anticlockwise until LED3 is

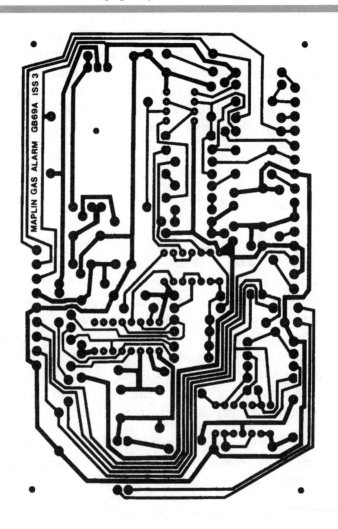

Figure 1.2 Track layout of main PCB

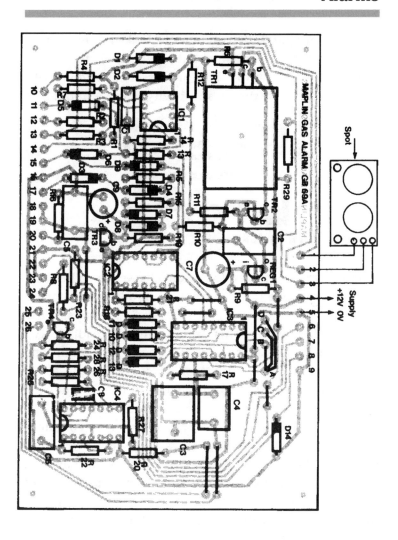

Figure 1.2   Continued

just on. Very carefully rotate RV1 clockwise again until the LED is just extinguished but can be made to light by pressing the CAL button. This process must be done with extreme care if maximum sensitivity is to be obtained.

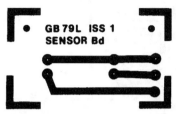

Figure 1.3   Gas sensor PCB

Disconnect the power for about 1 minute. When the power is reapplied, the *alarm LED* should light immediately but go out after a maximum of 30 seconds. If the LED remains on after this period, slightly readjust RV1 anticlockwise but make sure the above test conditions are met. Check that the alarm sounds when the *alarm LED* is alight and S3 is normal. The unit is now ready for use, but a further test may be carried out under actual working conditions. Place the sensor board in a container of about 5 litres capacity (e.g. a large ice cream container) and arrange a loose fitting card lid to cover it. Fill a small container of about 65 cc (e.g. a small aerosol lid) capacity with butane from an ordinary gas cigarette lighter and cover with card or a sheet of paper. With the alarm set, carefully slide the lid from the small container and pour the gas (butane is heavier than air) into the large container; then cover this container. The alarm should sound within a maximum of 6 minutes and remain latched until reset by S3. *Warning*, do not carry out this test near a naked flame, near incandescent material or when smoking!

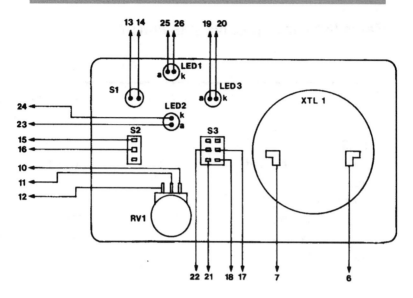

Figure 1.4   Wiring diagram

Installing the alarm

The sensor board may be located up to 5 metres from the main alarm unit. Most common explosive gases are heavier than air and therefore the sensors should be located at the lowest point where gas will collect. A free flow of air must be provided around the sensors and they must be kept free from contamination by oil or water. This alarm system is primarily designed for use in boats where power is supplied from a battery, but it could also be used in the home when fed from a suitable mains power supply (requiring about 800 mA at 12 volts). Finally, when gas is encountered, remember to ventilate the area well and beware of any form of ignition from naked flames, cigarettes, hot surfaces, sparks from electrical switching or other causes.

11

Parts list for explosive gas alarm

Resistors — All 0.6 W 1% metal film (unless specified

R1,3,7, 19,23	1 k min res	5	(M1K)
R2,4,5, 10,11,13, 15,20	4k7 min res	8	(M4K7)
R6	2k2 min res	1	(M2K2)
R8	47R min res	1	(M47R)
R9	10R min res	2	(M10R)
R12,21, 26,28	10 k min res	4	(M10K)
R14	100 k min res	1	(M100K)
R16	100R min res	2	(M100R)
R17	2M2 min res	1	(M2M2)
R18	2k7 min res	1	(M2K7)
R22,25	470 k min res	2	(M470K)
R24	68 k min res	1	(M68K)
R27	18 k min res	1	(M18K)
R29	330R min res	1	(M330R)
RV1	10 k pot lin	1	(FW02C)

Capacitors

C1	polylayer 0.01	1 (WW29G)
C2	polylayer 0.1	1 (WW41U)
C3	polylayer 1	1 (WW53H)
C4	polylayer 0.56	1 (WW50E)
C5,6	polylayer 0.22	2 (WW45Y)

C7	100 uF 25 V PC elect	1	(FF11M)
C8	330 ceramic	1	(WX62S)
C9	10 uF 100 V PC elect	1	(FF05F)

Semiconductors

D1,2	1N4001	2	(QL73Q)
D3-14	1N4148	12	(QL80B)
TR1	BD711	1	(WH15R)
TR2,3	BC327	2	(QB66W)
TR4	BC548	1	(QB73Q)
RG1	LM78L05ACZ	1	(QL26D)
IC1	CA3240E	1	(WQ21X)
IC2	HCF4011BEY	1	(QX05F)
IC3	HCF4060BEY	1	(QW40T)
IC4	HCF4001BEY	1	(QX01B)
LD1	LED yellow	1	(WL30H)
LD2	LED green	1	(WL28F)
LD3	LED red	1	(WL27E)

Miscellaneous

	5 mm LED clip	3	(YY40T)
S1	push switch	1	(FH59P)
S2	sub-min toggle A	1	(FH00A)
S3	sub-min toggle E	1	(FH04E)
	gas detector sensor	1	(FM87U)
	gas detector PCB	1	(GB69A)
	gas alarm sensor PCB	1	(GB79L)
	pin 2145	1	(FL24B)
	slotted heatsink	1	(FL58N)
	8-pin DIL socket	1	(BL17T)
	14-pin DIL socket	2	(BL18U)

	16-pin DIL socket	1	(BL19V)
XTL1	direct radiant piezo	1	(WF54J)
	10 M 7/0.2 wire blk	1	(BL00A)
	6BA 1/2 bolt	1	(EA06G)
	6BA nut	1	(EA16S)
	constructor guide	1	(XH79L)
	explosive gas alarm leaflet	1	(XT47B)

Optional

	case	1	(LH62S)
	K7B knob	1	(YX02C)
	fixings	as reqd	

The above parts (excluding optional) are available as a kit, order as LK60Q.

Peep alarm

The peep alarm is a light-operated sounder, simple but effective in its operation. In a dark environment, the unit remains quiet, and is for all practical purposes, switched off. However, when lights falls on the alarm's photosensor, the unit emits a loud, shrill tone. In addition, a light emitting diode (LED) illuminates, providing secondary indication that the circuit has been triggered. An on-board light sensitivity control is included in the design to allow operation in a variety of situations, over a range of light levels.

Circuit description

Figure 1.5 shows the circuit diagram of the module. The circuit is based around a 4093BE CMOS quad Schmitt NAND IC (IC1). Diode D1 prevents damage to the circuit if the battery is unintentionally connected the wrong way round. The photosensitive device used is an ORP12 light dependent resistor (LDR), chosen for its wide change in resistance relative to a comparatively small change in light level. Together with R1, R2 and RV1, LR1 forms a potential divider. As light falls on LR1, its resistance decreases, reducing the voltage at the centre of the divider chain. When the voltage level drops to the switching threshold of IC1a, the output of the gate switches from low to high. The output of IC1a is fed to IC1b, which forms a gated oscillator. When the logic level

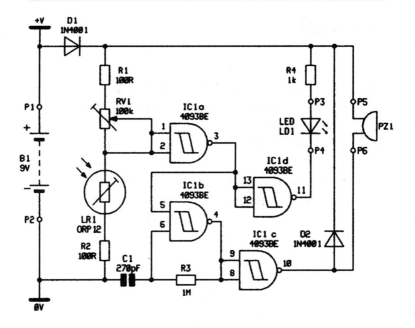

Figure 1.5 Circuit diagram

on pin 5 is high, the oscillator operates, but when the level is low, the oscillator is disabled. The operating frequency of the oscillator is determined by R3 and C1, the supplied values of R3 and C1 generating a frequency of approximately 4.6 kHz. IC1c forms an inverting buffer between the oscillator and sounder PZ1. Diode D2 is used to protect IC1 from any high voltage spikes which the sounder may produce. The output of IC1a is also fed to IC1d, which drives light emitting diode LD1. Resistor R4 limits the current through the LED.

Construction

The design uses a high quality, single-sided, glass fibre PCB, with a screenprinted legend, see Figure 1.6. Insert and solder the components onto the PCB, referring to the component legend, starting with resistors R1 to R4 and capacitor C1. Fit the IC socket, ensuring that the position of the notch at one end of the socket corresponds with that on the component legend. Do not fit the IC at this stage. P1 to P6, the PCB pins, are inserted

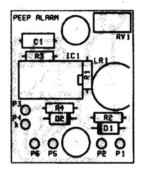

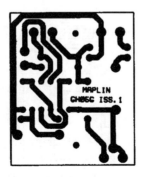

Figure 1.6   PCB legend and track

17

into the PCB from the track side using a hot soldering iron. The head of the pin should be heated to the extent that very little pressure is required to press it home into the PCB. After the pin has been inserted it may be soldered. Pre-set resistor RV1 is inserted so that its case corresponds with the outline on the component legend. When fitting D1 and D2, it is important that the correct polarity is observed; the negative end of the diode, indicated by a band at one end of the component, should be inserted so that it matches the band printed on the legend. Light dependent resistor LR1 should be mounted at a height of 10 mm above the PCB, as shown in Figure 1.7. IC1 may then be inserted into the socket. The IC should be positioned so that the notch at one end of the component corresponds with that in the socket. The battery clip is wired to P1 (red) and P2 (black). The light emitting diode, LD1, is soldered between P3 (anode) and P4 (cathode). The cathode of the LED is indicated by a flat edge on the side of the component body, and by the shorter of the two leads. PZ1 the piezo sounder, is wired between P5 and P6. All connections are illustrated in Figure 1.8, together with assembly information.

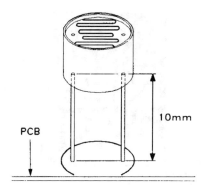

Figure 1.7 Mounting LR1

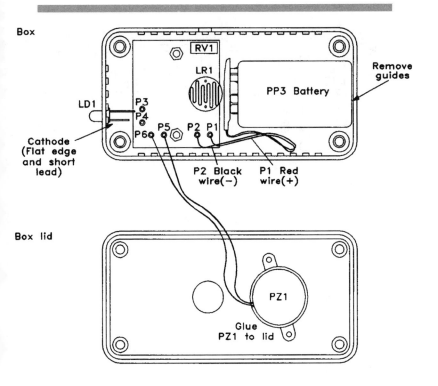

Figure 1.8 Assembly and wiring diagram

Housing

The peep alarm is housed in a small plastic box, which is included in the kit. The PCB is held in the box using M3 nuts and bolts, as shown in Figure 1.9. Additional M3 nuts are used as spacers under the PCB. The easiest method of assembly is to first insert the bolts into the underside of the box. Thread an M3 nut onto each of the bolts and screw down until locked into position. Now fit the PCB over the bolts, and lock it into place using a

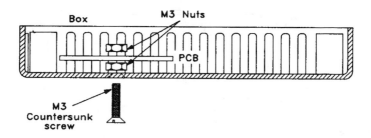

Figure 1.9   Mounting the PCB

second set of M3 nuts. It is necessary to remove some of the moulded plastic guides from the end of the box to allow the battery to fit comfortably. It is also necessary to drill several holes in the box. Apart from those used to mount the PCB, holes must be drilled in the case lid, so that light can fall on LR1, and to allow the buzzer to be heard as loudly as possible. The drilling details are shown in Figure 1.10. Piezo sounder PZ1 should be held in place on the inside of the box lid, using the epoxy adhesive supplied in the kit. This device should be pressed, as tightly as possible, against the lid until the adhesive sets. Note: the hole in the centre of the sounder must correspond with the hole in the box lid.

Testing

Before testing the module, it is a good idea to double-check your work to make sure that all of the components are inserted correctly, and to ensure there are no dry

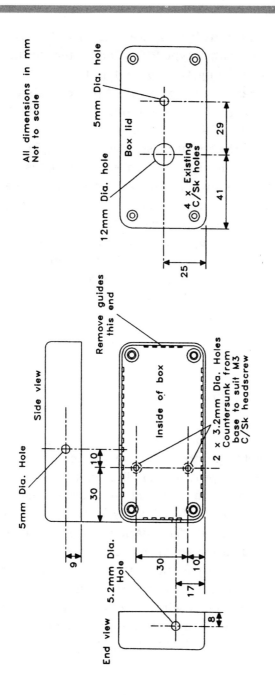

All dimensions in mm
Not to scale

5mm Dia. hole

Box lid

12mm Dia. hole

4 x Existing
C/Sk holes

29

41

25

Remove guides
this end

Side view

5mm Dia. Hole

Inside of box

2 x 3.2mm Dia. Holes
Countersunk from
base to suit M3
C/Sk headscrew

10

30

9

End view

5.2mm Dia.
Hole

30

10

17

8

Figure 1.10   Drilling details

21

joints, or solder bridges between tracks. If all is well, connect a 9 V alkaline type PP3-sized battery to the battery clip, observing the correct polarity. Place the module in a well-lit area, and adjust preset resistor RV1 until the sounder emits a loud high-pitched tone. If the unit is then placed in the dark (i.e. there is very little light falling on the sensitive surface of LR1), PZ1 should stop sounding. During the period when the unit is sounding, LD1 should glow, providing a secondary indication that the unit has been triggered. If you do not require LD1, this component may be omitted without any detrimental effect on the operation of the circuit. RV1 adjusts the sensitivity of the unit to light, and in practice it is necessary to adjust this control for optimum performance in any one particular situation. The circuit is set for maximum sensitivity when RV1 is fully anti-clockwise.

Using the module

The peep alarm may be used in a variety of applications requiring a light-operated alarm. In a typical application, it could be used to provide an indication that a cupboard or case is open; a situation that could arise from an attempt to gain unauthorised access, or simply by leaving it open unintentionally. The alarm is purely intended to give a local indication, and is not designed to be audible over a long distance. Obviously the alarm will only operate in an environment where there is a significant increase in light level when the cupboard or case is opened. The peep alarm is, however, designed to operate over a wide range of ambient light levels.

Another application for the peep alarm is in a dark-room, where it could be used to give warning when the door is opened, or a light is switched on.

In a dark environment, the circuit is in its quiescent state, and therefore draws very little current; in this condition, the unit should operate for many hours from an alkaline PP3-sized battery. When the unit is sounding, the current drain is several milliamps; for this reason it is recommended that the unit is not left in the triggered state for more that a few minutes at a time, otherwise the battery will become quickly drained.

Table 1.1 shows the specification of the prototype peep alarm; the figures stated may vary slightly in practice due to component tolerances.

Power supply	9 V alkaline type PP3-sized battery
Quiescent current drain	less than 1 μA
Operating current	14 mA
Alarm frequency	4.6 kHz approximately
PCB dimensions	41 x 34 mm approximately

Table 1.1   Specification of prototype

Peep alarm parts list

Resistors — All 0.6 W 1% metal film (unless specified)

R1,2	100 Ω	2	(M100R)
R3	1 M	1	(M1M)
R4	1 k	1	(M1K)
RV1	100 k vrt encl preset	1	(UH19V)

Capacitors

C1	270 pF 1% polystyrene	1	(BX50E)

Semiconductors

IC1	4093BE	1	(QW53H)
D1,2	1N4001	2	(QL73Q)
LD1	2 mA 5 mm LED red	1	(UK48C)

Miscellaneous

PZ1	low profile sounder	1	(KU57M)
LR1	LDR ORP12	1	(HB10L)
	PP3 clip	1	(HF28F)
	pin 2145	1	(FL24B)
	14-pin DIL socket	1	(BL18U)
	PCB	1	(GH06G)
	2002 ABS box	1	(WY03D)
	M3 10 mm poziscrew	1	(LR57M)
	M3 steel nut	1	(JD61R)
	double bubble sachet	1	(FL45Y)
	instruction leaflet	1	(XT41U)
	constructors' guide	1	(XH79L)

A kit of the above parts is available order as LP50E

Anti-theft device

The design for this anti-theft device (ATD) came about because of a story that I know will be familiar to many readers. After spending all of Sunday morning at the local casualty department with my wife, who had broken her wrist, we returned home at lunch-time to find that our house had been broken into.

The err... *intruder*, for lack of a printable description, had gone round the back of the house, forced a kitchen window open, then climbed through leaving a trail of muddy footprints all over the worktops and floor. By the time he had got to the dining room door all hell must have broken loose as the PIR sensor picked him up and set off the burglar alarm. The two sounders in the kitchen, the two in the lounge, plus the two upstairs and the external siren all going off in unison. It must have been a shock to his system, because he didn't get any further, the muddy footprints showed it was a quick exit back the way he came. All this and he didn't get any swag. One up for the good guys. Of course the police were called, but we became just another statistic and nothing has been heard of the matter since. For those of you who I can still hear muttering those immortal words "It will never happen to me," I say "Don't be so complacent — after all, it happened to me!"

Paranoia

I know the alarm worked and scared him off, but you start getting paranoid and asking yourself questions.

Like: "What if I hadn't switched the alarm on, or for some reason it had not worked?" "What if he had just grabbed the microwave or portable TV from the breakfast bar and ran off, what good would the burglar alarm had been then?" Some sort of back-up system was required.

The cure

What was needed was a small device that could be hidden away inside almost any domestic appliance or piece of furniture such as a TV, video, hi-fi, microwave or computer and just lay there dormant for two or three years. If the host appliance was moved a loud screeching sound would be emitted which would go on and on until either the battery ran out or the alarm was switched off. No one would go walking down the street with a screaming microwave under their arm, or would they? The alarm would have to be able to be switched off from outside the appliance and as I didn't fancy drilling any holes for an on-off switch in my new video recorder, some other method of switching was required. It would have to be simple to make and last but most importantly because of the trace of Scottish blood in my veins, cheap.

A case of something for nothing

Usually most home-made projects, no matter how good the electronics, suffer from the same old problem. That is, what the finished article looks like. Most attractive housings are usually expensive, the cost can easily be two or three times that of the electronics that goes in them. As most individuals can't afford the thousands of pounds involved in getting an injection moulded case made for their Mk. V micro-controlled egg timer, most projects end up being housed in standard off the shelf plastic or metal boxes. All the holes for controls etc., have to be drilled by hand and any legends have to be put on with transfer letters or stamped tape. It's all too easy to end up with something that — well how can I put this? — looks home-made. A stick-on overlay can cover a multitude of sins and turn even the humblest of boxes into something quite presentable. But this is still an expensive option for an individual who might only want to build two or three egg timers. Even for large firms like Maplin who want to sell hundreds or better still, thousands of their designs, an overlay would still add to the overall price. So how can Maplin stock a custom designed injection moulded case for this project and provide a pair of piezo sounders (ready fitted), two output transformers and two driver transistors for only £3.75?

Full of eastern promise

The answer is by using the case and other components from a twin-unit piezo siren which has been mass produced somewhere in the far east. A new PCB has been designed that contains a sensitive movement detector together with latch, a swept frequency audio oscillator and transformer output stage to drive two piezo sounders. A wide range sensitivity control together with

Home security projects

controls to maximise the sound output are also provided on board. Enough room is left in the case for a 9 V PP3-sized battery which will power the ATD in its passive state for three to four years, and up to 10 hours when activated. A small magnetic switch of the type used for door contacts in burglar alarms is connected to the main box by a short length of cable. This allows the main box to be placed in any convenient spot inside the appliance. The magnetic switch can be located elsewhere so that a magnet can be wiped over the outside of the appliance to switch the alarm off, or stop the alarm from sounding whilst the rightful owner moves it.

Circuit description

The circuit for the ATD can be split up into three distinct sections: these are shown in the block diagram in Figure 1.11, and in the circuit diagram of Figure 1.12.

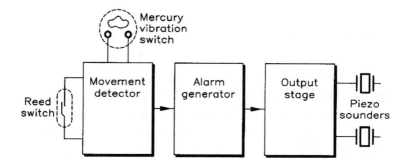

Figure 1.11   The ATD block diagram

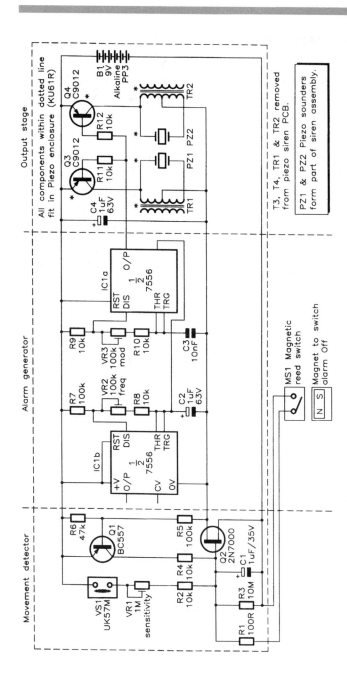

Figure 1.12 The ATD circuit diagram

Home security projects

Movement detector

This is based around mercury vibration switch VS1. No matter what the orientation of VS1, its contacts are always open, while it is stationary. As VS1 is moved, its contacts open and close, producing a stream of short pulses (see Figure 1.13). Provided the magnetic switch MS1 is open, each pulse will cause C1 to acquire some more charge. VR1 is the sensitivity control and the higher its resistance the more pulses are needed to charge C1. Due to the very low gate current (10 nA) of the *Fetlington* (FET Darlington) transistor Q2, its presence does not affect the charging of C1. As the voltage on C1 rises to approximately +1 V, Q2 starts to turn on. As the voltage on its drain starts to fall the voltage across resistors R5 and R6 rises, turning on transistor Q1. R4 is then pulled high, quickly charging C1. Once latched in this state, power is applied to the rest of the circuit. If the magnetic switch MS1 is closed the ratio of R1 to R2, VR1 and R4 ensures that Q2 is turned off even if VS1 is closed. The purpose of R1 is to discharge C1 when no charge pulses are being generated, this ensures that a number of short duration knocks over a long period of time will not activate the ATD.

Alarm waveform generator

The alarm waveform consists of an audio frequency squarewave being swept up and down in frequency by a low frequency modulation waveform. This is achieved by using IC1, a 7556 dual CMOS timer. IC1b, R7, R8, VR2

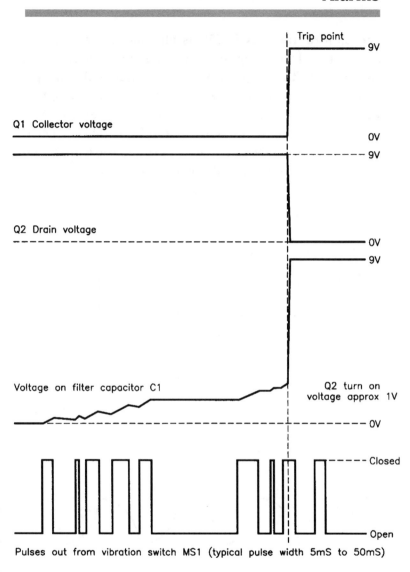

Trip point

9V

Q1 Collector voltage

0V

9V

Q2 Drain voltage

0V

9V

Voltage on filter capacitor C1

Q2 turn on
voltage approx 1V

0V

Closed

Open

Pulses out from vibration switch MS1 (typical pulse width 5mS to 50mS)

Figure 1.13   Movement detector, typical waveforms

31

and C2 form the classic 555 astable multivibrator. The modulation frequency may be adjusted from approximately 6 Hz to 50 Hz by VR2. The charge/discharge voltage on pins 1 and 5, see Figure 1.14, is applied to the voltage control pin (pin 10) of the other astable multivibrator (IC1a, R9, R10, C3 and VR3). This sweeps the audio frequency up and down around the centre frequency which is set by VR3 (600 Hz to 5 kHz).

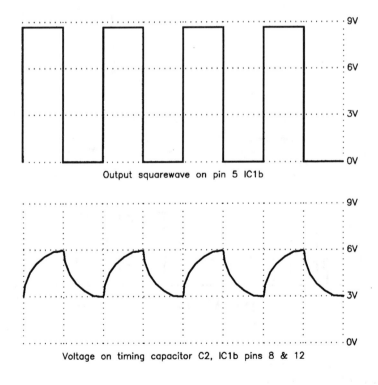

Output squarewave on pin 5 IC1b

Voltage on timing capacitor C2, IC1b pins 8 & 12

Figure 1.14 Modulation waveform

Output stage

The output stage consists of two identical circuits (R11, T3, TR1 and R12, T4, TR2) each driving a piezo sounder (PZ1 and PZ2). The output transistor is driven on (0 V) and off (+9 V) by the squarewave output (pin 5) of IC1b. The centre tap of the transformer is connected to the switched 0 V line, and one end is connected to the collector of the output transistor.

As the other end is left floating, transformer action dictates that 18 V peak-to-peak (twice the drive voltage) will be developed across the two outer ends of the transformer. Each time the drive transistor switches off a voltage is produced, due to the inductance of the transformer. With no load connected to the transformer this manifests itself as a negative going spike, which can vary between 50 to 150 V depending on frequency. This brings the average voltage across the sounder to approximately 40 to 50 V peak-to-peak, which accounts for the high level of sound generated by the unit.

Cannibalism

The first thing that needs to be done is to remove the four cross-head screws from the back of the twin-unit piezo sounder and remove its cover. The PCB should then just drop out, if not, ease it out with a small screwdriver. Desolder the four wires from the PCB that go to the sounders.

Home security projects

If the PCB from your sounder has the reference number S100(A), then the components to remove are marked on the board as Q3, Q4, T1 and T2. Be careful when desoldering the transformers, as the leads are quite fragile. Make sure that all the leads are free from the pads before pulling them out of the board (see Desoldering Tips).

If you have a different PCB then you will have to trace the tracks from the transformer to its transistor to find the ones to remove. If the transistors you remove do not have the type numbers A1270 or C9120, then to make sure it is a PNP type, check that its emitter (see Figure 1.15), goes to the positive rail. It in doubt then refer to the Parts List for an alternative.

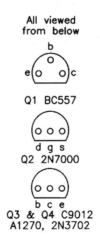

Figure 1.15   Transistor bases

First things first

Construction of the ATD is relatively straightforward, and does not require the use of specialist tools or test equipment, but before assembling the PCB a few things should be done. Firstly, the backplate as supplied has two triangular shaped mounting brackets on it. These need cutting off with either a sharp craft knife (be careful), or a junior hacksaw.

Next, a hole needs to be drilled in the backplate to allow access to the sensitivity control. The size is not too important, approximately 5 mm diameter, and Figure 1.16 shows its position. Last of all remove the five screws from the back of the magnetic switch MS1, and carefully

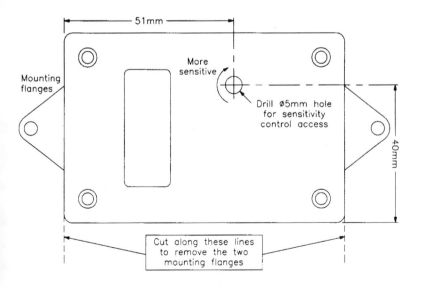

Figure 1.16 Backplate drawing

bend up the wires coming from the reed switch. Cut out three nicks in the back of MS1 for the cable to sit in as shown in Figure 1.18. Solder a 1 m length of zip connecting cable (*figure of eight* type) to the wires and push the cable into the nicks. Before going on to the next stage strip the other end of the cable and check that the switch still works. Mix up some 5-minute epoxy resin and fill the back of the switch with it, ensuring the cable stays in place. Wipe off any excess glue with a damp cloth and leave to set.

Constructing the PCB

The circuit is constructed on a PCB to achieve a compact and reliable circuit. Using the track layout diagram Figure 1.17 for reference, start by inserting and soldering links LK1 and LK2, followed by resistors R1 to R12 and preset resistors VR1 to VR3. Capacitors C1 to C4 can then be fitted, observing the correct polarity of the three 1 μF electrolytics. The negative lead is denoted by negative signs on the case and should be inserted away from the positive (+) sign shown in Figure 1.17.

Next insert and solder the four transistors Q1 to T4, ensuring that the transistor cases match the outlines in Figure 1.17. Then, ensuring you have taken the usual precautions against static damage, insert and solder IC1 into the PCB aligning the notch denoting pin 1 as shown in Figure 1.17. Transformers TR1 and TR2 may then be soldered into the PCB (see notes on transformers).

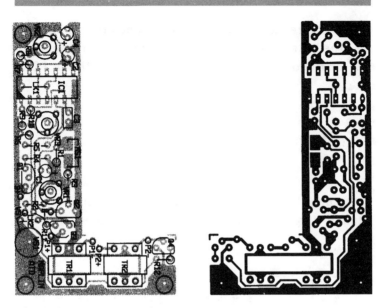

Figure 1.17   PCB legend and track

Vibration switch VS1 can then be inserted and soldered. Solder a short length of wire between the outer case of VS1 and the hole marked VS in Figure 1.17. Next, strip back the insulation on the PP3 battery clip so that there is approximately 60 mm of red and 50 mm of black wire left, insert and solder into the holes marked B+ and B– respectively. Because there is no diode to protect against battery reversal, check that the red wire is the one nearest the longest edge of the PCB.

Before going any further check that all the components are fitted correctly and that there are no dry joints or solder bridges on the track side of the board. Using a PCB cleaning solution (DM83E) remove all the flux from the PCB. Using a small screwdriver centre the wipers of presets VR1, VR2 and VR3.

Home security projects

Solder the red and black wires from one of the sounders into the holes P1+ and P1– respectfully, repeat the process for the other sounder using the holes P2+ and P2–.

Transformer identification

The twin-unit piezo siren design is not likely to change in the near future. Therefore, these tests should only be needed if you are building the project some time after

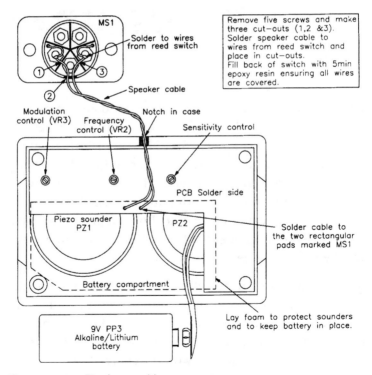

Figure 1.18   Final assembly

the publication date, or you are using a siren from a different source. It is a good idea to check the transformers anyway, just in case one has been damaged during removal from the siren PCB.

There are number of similar sirens on the market (at least 4) and the design of the transformer varies from manufacturer to manufacturer. All the ones tested so far use a centre tapped 20 Ω–0–20 Ω winding (approximately) to drive the sounders. Some simple tests have to be done to find this winding, Figure 9(a) shows the primary and secondary winding connections. Once found the transformers can be put into one of two groups (A or B), this determines which way round they are to be inserted into the PCB.

Measure the resistance between the pairs of pins shown in Table 1.2. Compare your results with those in the table to determine if it is a type A or type B transformer. Figures 1.19(b) and 1.19(c) show which way round the transformers have to be inserted into the PCB. The PCB legend shows the orientation of a type A transformer, as found in sounders with the variant number S100(A) on the PCB.

	Resistance between pins				
1 and 2	2 and 3	2 and 4	2 and 5	4 and 5	Type
42 Ω	42 Ω	21 Ω	21 Ω	42 Ω	A
21 Ω	21 Ω	NC	NC	2 Ω	B
24 Ω	24 Ω	NC	NC	45 Ω	B

All resistances are approximate

Table 1.2 Transformer identification

Finishing off

Once again double-check that all the components are fitted correctly and that there are no dry joints or short circuits, then referring to Figure 1.18 for guidance, the finished PCB can be fitted into the box. Carefully tuck the wires from the sounders under the PCB. Fit the piece of foam into the bottom of the battery compartment. Finally, the wires from the remote magnetic switch (MS1) can be soldered to the two rectangular pads on the track side of the PCB and taken out of the notch in the edge of the case.

Testing

Put a magnet against MS1 then connect the battery, if possible through a current meter. The ATD should remain silent and less than 50 nA (0.05 μA) leakage current should be flowing. If the ATD is shaken the current consumption will rise. If it rises to more than 1 mA there is a fault.

Remove the current meter and the magnet from MS1 and shake the ATD, it should start sounding. Adjust the frequency control (VR3) until maximum volume is achieved (at approximately 2.8 kHz). The modulation control (VR2) can be adjusted to create the most disturbing noise possible. Returning the magnet to MS1 should silence the alarm. The sensitivity control (VR1) can be turned clockwise to make it more sensitive and anti-clockwise to become less sensitive. When the ATD is active measure the current consumption once again, it should be in

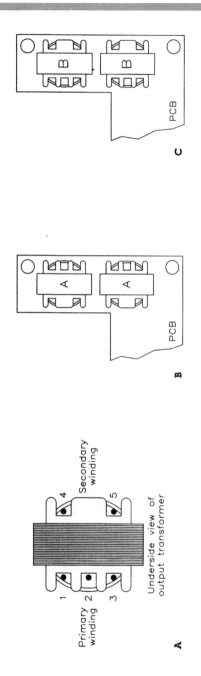

Figure 1.19 (a) Underside view of output transformer, (b) type A orientation, (c) type B orientation

41

the region of 50 mA. Also check that the voltage across the drain and source of the 2N7000 is less than 200 mV. Once you are happy that all is well the back can be put on, making sure the hole lines up with the sensitivity control.

It is a good idea to do a long-term test on the ATD to be a hundred per cent sure that it won't just start and then pack up half an hour later because of a faulty component. Start it sounding and then bury it in the linen basket or wrap it in towels, anything to muffle the sound and just let it run for a couple of hours or so, checking occasionally.

Warning about guarantees

If the TV or video that you want to fit the ATD into is rented, then you will have to get the permission of the rental company before you do so. Most guarantees will be invalidated if covers are removed and alien objects inserted. Because of the chance of fatal electric shock *always* switch the appliance *off* and remove it from the mains supply before removing any covers. Be especially careful when working on TVs because the high-voltage section can still have capacitors charged up to many thousands of volts long after the power has been removed. *You have been warned!!*

Fitting the beast

Three or four double-sided sticky pads will usually secure the ATD to the inside of its host, but make sure you brush off any dust and clean the surface before doing so. The magnetic switch (MS1) can be secured by one sticky pad so that a magnet can be easily wiped over from the outside. When looking for a suitable place to fit the ATD try to avoid sources of heat such as large heatsinks and transformers. Keep well away from any moving parts and if fitting inside a TV, don't fit the main box above the tube in case it falls off and damages the tube, and keep away from the high-voltage section. Also make sure that when replacing any covers they do not foul on the ATD or the remote wire and switch.

The sensitivity control will need adjusting to suit each installation and will be a compromise between sensitivity to movement and the odd knock with the vacuum cleaner etc. It is a good idea to mark the location of the magnetic switch on the outside of the case with a small sticky label or similar, so that if the ATD gets woken up in a year's time you will know where to put the magnet. Don't forget where you put the magnet either! From experience, it is a good idea to keep a record of where the magnetic switches are in each appliance, and also when the battery needs replacing. Also checks should be made every month or so to make sure it is doing the job it was designed for.

Not all batteries are equal

The ATD has been designed to work off a standard alkaline 9 V PP3-sized battery which has a 500 mAh or more capacity. As the shelf life of these batteries is quoted as

Home security projects

5 years and the ATD takes virtually no current when off, it would seem that in theory an ATD could lie dormant in its host for 5 years and still work when woken up. In practice replacing the battery every 3 years would be a wise move.

Under no circumstances use a zinc-carbon or zinc-chloride type battery as they do not have as long a shelf life or the capacity of an alkaline battery, and are more likely to leak.

Moving house

If something that has an ATD installed has to be moved a long distance, i.e. moving house, use masking tape to hold a magnet over the concealed switch. Even a five hour journey will only use up a maximum of 1% of an alkaline PP3's 500 mA capacity. So as you can see this won't have much impact on battery life.

Extra security

If you are worried that some of the criminal fraternity read this magazine to keep abreast of the latest technology that's trying to put them out of business, and that they will now go to work armed with a couple of magnets, then you can wire up two magnetic switches in series. This will make disarming the ATD almost impossible for anyone who does not know both their positions, especially in the short time a burglar wants to spend in your house.

Note: the circuit and information presented here must be considered as a basis for your own experimentation. No warranty is given or implied for suitability in particular applications — Maplin cannot support this information in any way. However, where possible, we endeavour to check that information is correct and that circuits will function as stated.

Desoldering tips

Tip 1. When using desolder braid to remove components, check that the braid has not passed its *sell by date* by seeing if it will remove solder from a newly soldered joint. If it won't then the odds are against it removing solder from an old oxidised joint, One way to rejuvenate old braid is to run some new solder onto it with a soldering iron and let it wick up for a centimetre or so. Cut off all but a millimetre of the newly soldered braid. Desolder joints using this end and keep trimming it off as required.

Tip 2. When using a desolder pump always keep the nozzle clear and check that it is working by pressing the nozzle hard into the palm of your hand. If the plunger does not move a lot slower than normal and you cannot feel any suction then its not working properly. When desoldering, heat the joint up with the iron, keeping the nozzle as close as possible. At the last second remove the iron from the joint, and then press the button. This might take a bit of practice to get right but it will stop the iron from damaging the tracks on PCBs.

Home security projects

Tip 3. Whatever method you choose to desolder components, always add some new solder to old joints to make them shiny again. The flux in the new solder makes the joint more fluid when heated and a lot easier to desolder.

Anti-theft device parts list

Resistors — All $^1/_8$ W carbon film (unless specified)

R1	100 Ω	1	(U100R)
R2,4,			
8–12	10 k	7	(U10K)
R3	10 M	1	(B10M)
R5,7	100 k	2	(U100K)
R6	47 k	1	(U47K)
VR1	1 M cermet preset	1	(WR45Y)
VR2,3	100 k cermet preset	2	(WR44X)

Capacitors

C1,2,4	1 µF 63 V sub-min electrolytic	3	(YY31J)
C3	10 nF monolithic ceramic	1	(RA44X)

Semiconductors

Q1	BC557	1	(QQ16S)
Q2	2N7000 Fetlington (FET Darlington)	1	(UF89W)
Q3,4	2N3702, (A1270 or C9120 type)	2	(QR36D)
IC1	ICM7556 CMOS timer	1	(CP96E)

Miscellaneous

VS1	vibration switch	1	(UK57M)
MS1	magnetic reed switch	1	(JU65V)

Home security projects

PS1	twin piezo sounder	1	(KU61R)
BC1	PP3-sized battery clip	1	(HF28F)
B1	9 V PP3-sized battery alkaline	1	(JY49D)
	zip connecting cable	1	(XR39N)
	self-adhesive pads	1	(HB22Y)
	PCB single sided (see text)	1	(HX01B)
	constructors' guide	1	(XH79L)

Home alarm

The need for an inexpensive but effective home alarm is becoming of greater importance as time passes. This simple unit has several features which are normally found on more expensive alarms now available.

No alarm can offer complete protection against the determined professional burglar, but it will act as a strong deterrent to the small-time thief. This alarm can be triggered from a wide range of security sensors, such as a reed switch, pressure mat, window foil, etc.

It can be expanded further to include heat, vibration and infra-red detectors. This system flexibility ensures that the potential thief never knows just what to expect from house to house.

The operating principle is that when the arm key is turned on, you then have up to 60 seconds to secure the building; during this exit time a buzzer will be sounding. After this delay the alarm is active and ready to be triggered by any of the sensors in the security loops. If you are still inside the building a red indicator acts as a reminder that the system is active.

When one or more of the sensors is disturbed the alarm is triggered, lighting the intruder indicator and sounding the buzzer. You now have up to 60 seconds of entry delay to deactivate the alarm before the main siren goes off.

The alarm is equipped with two additional security loops which are permanently active and immediately trigger

Home security projects

the main siren when disturbed. The first of these is a *tamper-proof* loop which when broken sounds the siren until it is reconnected. The second loop is used for manually operated panic buttons, and if either loop is triggered, the help indicator will light up until the alarm is reset by the key switch being turned from the standby to the armed condition.

Circuit description

A circuit diagram detailing the complete unit is shown in Figure 1.20. The timing and control of the sequence of events is governed by the logical function of the following integrated circuits (IC) and transistors (TR):

- IC1   TLC555   timer, 1 to 60 seconds exit/entry delay,

- IC2a,b   4001BE   latch, alarm active and exit/entry delay select,

- IC2c,d   4001BE   latch, security loop trigger,

- IC3a,b   4011BE   gate, siren entry delay, /

- IC3c,d   4011BE   latch, tamper and panic loop trigger,

- TR1   BC548   timer entry retrigger,

- TR2   BC548   exit/entry buzzer switch,

- TR3   TIP122   siren switch.

PCB construction

Removal of a misplaced component will be fairly difficult so please double-check each component type, value,
and its polarity where appropriate, before soldering! The
printed circuit board (PCB) has a legend to help you
correctly position each item, see Figure 1.21. Install all
the components including the buzzer which is secured
to the board using the 8BA screws and nuts. When mounting the red light emitting diodes (LEDs) you must ensure
that their polarity and positioning is correct. The cathode (K) is denoted by the shorter of the two leads and
by a flat on the bottom edge of the package as shown in
Figure 1.22.

When the assembly of the circuit board is complete you
should check your work very carefully, making sure that
all solder joints are sound. It is also very important that
the solder (track) side of the PCB does not have any
trimmed component leads standing proud by more than
2 mm, as they may cause short circuits.

Testing

All the tests can be made with the minimum of equipment. You will need a regulated 12 V d.c. power supply,
or 12 V battery, capable of providing the current required
by the siren you have chosen to use, but which must not
draw more than 1 A, and some hook-up wire. The following quoted meter readings were taken from the prototype
using a digital multimeter; some of the readings you obtain may vary slightly depending upon the type of meter
you use.

Home security projects

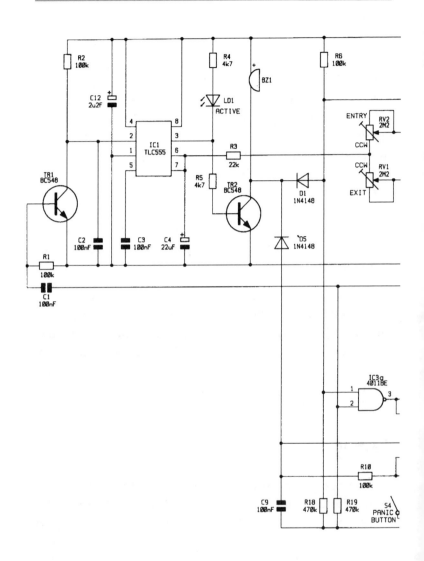

Figure 1.20 Circuit diagram

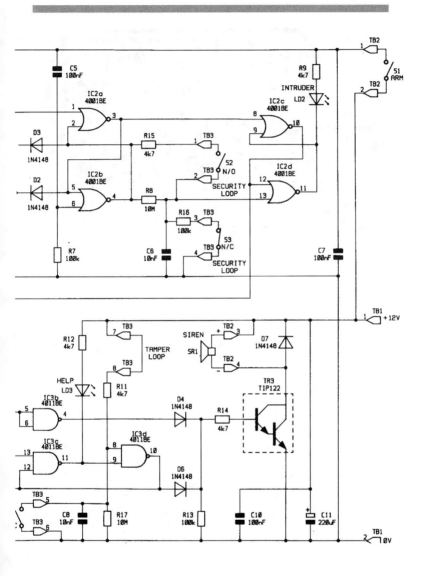

Figure 1.20   Continued

Home security projects

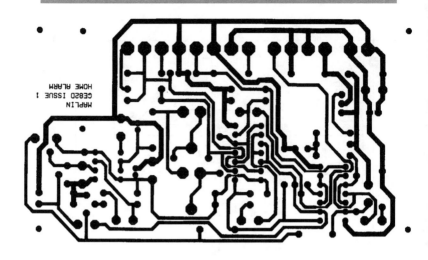

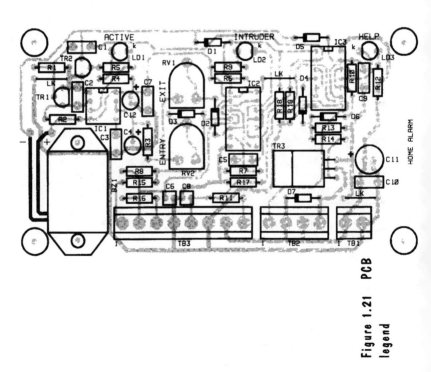

Figure 1.21 PCB
legend

54

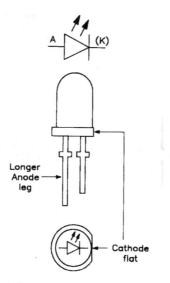

Figure 1.22 LED information

Before you commence testing the unit, set the two pre-sets, RV1 and RV2, fully counter-clockwise (CCW). Next, fit two wire links to terminal block 3 (TB3) as follows:

● terminal 3 to 4 (normally closed security loop), see Figure 1.23,

● terminal 7 to 8 (normally closed tamper loop).

Prepare the key switch S1 and siren SR1, then connect them to TB2 as follows:

● S1 to terminals 1 and 2 (set key switch to its clock-wise *on* position),

● SR1 red lead (+V) to terminal 3 and black (–V) to 4.

Do not connect any power to TB1 until it is called for during the testing procedure!

55

Home security projects

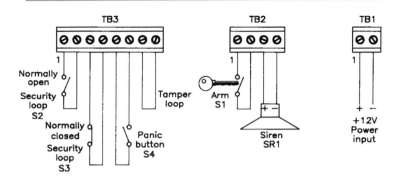

Figure 1.23   Wiring diagram

The first test is to ensure that there are no short circuits before you connect the power supply. Set your multimeter to read *ohms* (Ω) on its 20 kΩ or equivalent resistance range, and connect the test probes to terminals 1 and 2 of TB1. With the probes either way round a reading greater than 1 kΩ should be obtained.

Before testing the current consumption of the unit, set S1 to its counter-clockwise *off* position. It is also recommended that the output of the siren is muffled to reduce its sound intensity! Now set your meter to read d.c. mA and place it in series with the positive line of the power supply to TB1 terminal 1. When the negative power line is connected to TB1 terminal 2 the following should be observed:

● none of the LEDs should light,

● buzzer should not sound,

● siren should not sound,

● a current reading of approximately 1.3 µA should be obtained.

Alarms

Next, turn the key switch clockwise to its *on* position, which should arm the alarm system. At the same time the buzzer should sound for approximately one second, and at the end of this time the active indicator, LD1, should light. Repeat this procedure whilst advancing the position of RV1 in small steps. Each time the setting is increased, the exit delay should also increase until, when RV1 is fully clockwise, a maximum delay of approximately 60 seconds is reached. During the exit delay a current reading of approximately 35 mA should be observed, dropping down to 3 mA when the system is active.

To test the alarm trigger function, each security loop on TB3 must be individually opened or closed. Set RV1 and RV2 fully counter-clockwise. Then turn the key switch S1 to its *on* position (system armed) and perform the following tests:

1a   link terminals 1 and 2 (normally open security loop).

Result — active indicator LD1 goes out. Intruder indicator LD2 lights up. Buzzer sounds for a one second entry delay. Active indicator LD1 lights up. Siren SR1 sounds. Supply current reading increases by that demanded by your siren.

1b   reset the alarm by turning the key switch off (system standby).

Result — all indicators go out. Siren stops sounding. Current reading drops back to 1.3 μA. Remove the link from terminals 1 and 2.

2a   re-arm the system. Remove the link from terminals 3 and 4 (normally closed security loop).

Home security projects

Result — same as test 1a.

2b   reset the alarm (system standby).

Result — same as test 1b. Refit the link at terminals 3 and 4.

3a   link terminals 5 and 6 (normally open panic loop).

Result — help/tamper indicator LD3 lights up. Siren sounds immediately.

3b   remove the link from terminals 5 and 6. Arm the system.

Result — help/tamper indicator LD3 goes out. Siren stops sounding. Reset the alarm (system standby).

4a   remove the link from terminals 7 and 8 (normally closed tamper loop).

Result — same as test 3a.

4b   refit the link at terminals 7 and 8. Arm the system.

Result — same as test 3b. Reset the alarm (system standby).

This completes the testing of the alarm. Now disconnect the power, test links and your multimeter from the unit.

Using the alarm

Before the completed module can be used in a practical working environment the following operating conditions should be considered.

Power supply

As it can be seen from Figure 1.24, there are three basic power supply options. The simplest is shown in (a) where a 12 V battery supplies all the power to the alarm system. It must be remembered that the capacity of the battery will determine the effective operational life of the unit, which is in turn governed by the sequence of events. In its standby condition the alarm draws very little current (1.3 μA), so even small capacity batteries will last for a relatively long period. However, when armed the supply current increases to 3 mA and once triggered the current drawn by the siren can go up to 1 A depending upon the type you have selected. For this reason it is advantageous to use a battery with a high capacity to ensure good long term operation.

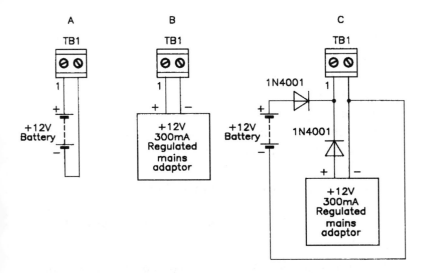

Figure 1.24 Power supplies

The second option, (b), uses a 12 V regulated mains adapter which must have sufficient current capacity to drive the siren. The prototype used a siren which drew less than 300 mA, so only a small mains adapter (YB23A) was necessary. Although this option gives a virtually uninterrupted supply you must take into account the fact that should the 240 V a.c. mains fail, then the alarm system is left inoperable. For this reason a secondary battery back-up supply is advantageous, as indicated in option (c) of Figure 1.24. Combining both supplies requires two 1N4001 1 A rectifier diodes (QL73Q) connected as shown which effectively isolate the supplies from each other.

Siren

There is a wide range of sirens available from Maplin, and their current consumption ranges from 150 mA (JK42V) to 300 mA (Y2030). As the alarm has a switching capacity of 1 A it is permissible to use more than one siren as long as the total current demand is less than 1 A and within the capacity of the power supply, in other words the following combinations are possible.

Where all sirens are wired in parallel:

● up to 6 150 mA micro sirens (JK42V), total current 900 mA,

● up to 3 300 mA miniature piezo sirens (JK43W), total current 900 mA,

- up to 3 300 mA low cost electronic sirens (YK60Q), total current 900 mA,

- 1 600 mA metal horn siren (YN59P).

Or any combination of these totalling 1 A.

The siren chosen for testing the prototype was the micro siren (JK42V) drawing only 150 mA but producing a very loud output. All the sirens are loud and there must be careful consideration when choosing where to install them as you could very easily upset the neighbours!

Box

It's good practice to build as much as possible into the box, because the more wiring there is outside the easier it is to tamper with. The choice of box will depend upon the following design criteria:

- size of assembled alarm PCB,
- size of batteries,
- size of mains power supply. Can be external if battery back-up fitted,
- size of siren. Additional sirens can be outside the box,
- box material plastic or metal,
- free-standing or wall mounted,
- front panel layout and markings.

A comprehensive range of boxes is available from Maplin which can be found in the current catalogue.

Security sensors

Your selection of sensors to use with the home alarm will depend on your requirements. Here is a list to help you select those most suitable for your needs.

Reed switches

Recessed door or window	YW46A
Recessed five terminal door or window	FK77J
Recessed panel pin fixing, five terminal door or window	JU65V
Surface mounting, door or window	YW47B

Panic buttons

Round panic button	FK46A
Help button	FP12N
Metal panic button	YZ67X

Pressure mats

Standard carpet	YB91Y
Stair carpet	FK79L

Window protection

Window foil	YW50E
Foil terminals	YW51F
Glass break detector	FP11M

Infra-red

Beam break detector	BZ64U
Hi immunity PIR	AG81C
Pulse count PIR	AJ96

Miscellaneous

Heat detector	FK47B
Vibration detector	FK78K
Door junction box	YW48C
5-way junction box	YW49D
8-way junction box	FK76H
4-core burglar alarm cable (1 metre)	XR89W
(100 metres)	PA77J

Power supply input voltage:	6 V to 12 V d.c.
Current at 12 V:	Standby = 1.3 µA
	Active = 3 mA
	Intruder = 7.5 mA
	Help = 4.5 mA
Siren switching current:	1 A maximum
Exit/entry delay time:	1 to 60 seconds
Security loops:	Normally open and normally closed
Tamper loop:	Normally closed
Panic button loop:	Normally open

Table 1.3   Specifications of prototype

Low cost home alarm parts list

Resistors — All 0.6W 1% metal film

R1,2,6,7,			
10,13,16	100 k	7	(M100K)
R3	22 k	1	(M22K)
R4,5,9,11,			
12,14,15	4k7	7	(M4K7)
R8,17	10 M	2	(M10M)
R18,19	470 k	2	(M470K)
RV1,2	2M2 hor encl preset	2	(UH10L)

Capacitors

C1,2,3,5,			
7,9,10	100 nF disc ceramic	7	(YR75S)
C4	22 µF 25 V PC electrolytic	1	(FF06G)
C6,8	10 nF disc ceramic	2	(BX00A)
C11	220 µF 16 V PC electrolytic	1	(FF13P)
C12	2µ2F 100 V PC electrolytic	1	(FF02C)

Semiconductors

D1,2,3,			
4,5,6,7	1N4148	7	(QL80B)
LD1,2,3	2 mA 5 mm LED red	3	(UK48C)
TR1,2	BC548	2	(QB73Q)
TR3	TIP122	1	(WQ73Q)
IC1	TLC555CP	1	(RA76H)
1C2	4001BE	1	(QX01B)
1C3	4011BE	1	(QX05F)

Miscellaneous

	8-pin DIL socket	1	(BL17T)
	14-pin DIL socket	2	(BL18U)
	low cost alarm PCB	1	(GE82D)
TB1	2-way PC terminal	1	(FT38R)
TB2	4-way PC terminal	1	(RK73Q)
TB3	8-way PC terminal	1	(RK38R)
S1	plas key switch	1	(FV42V)
BZ1	12 V buzzer	1	(FL40T)
	8BA $^1/_2$in bolt	1	(BF09K)
	8BA nut	1	(BF19V)
	constructors' guide	1	(XH79L)
	home alarm leaflet	1	(XT03D)

Optional

SR1	micro piezo siren	1	(JK42V)
S4	help button	as reqd	(FP12N)

Security loop switches

S2	std pressure mat	as reqd	(YB91Y)
S3	surface BA reed	as reqd	(YW47B)
	window foil	as reqd	(YW50E)
	foil terms	as reqd	(YW51F)
	4-wire burglar cable	as reqd	(XR89W)

Batteries

	AA alkaline	8	(ZB47B)
	12 V battery box	1	(RK44K)

Home security projects

PP3-sized battery clip	1	(HF28F)
918S battery	1	(ZB66W)
HP992 general purpose	2	(YJ23A)
a.c. adaptor regulated	1	(YB23A)
1N4001 rectifier diode	2	(QL73Q)

Loop alarm

This little gadget was developed in response to a real problem — the pilfering of items from a shop. During peak shopping periods like Saturdays, shop staff are usually too preoccupied with customers to notice petty thieves lift items of often quite significant value (personal cassette players and cameras, for example), and then disappear undetected. Of course, this problem — thanks to a severe recession and worsening unemployment situation — is increasing steadily and so shopkeepers need to invest in some kind of security system, to protect their stock. In the case of the simple but effective Loop Alarm described here, this investment need not be too great.

Householders are also vulnerable — leaving aside obvious targets like bicycles, surprisingly large objects like ladders (and in one case, a satellite dish with a diameter of six feet!), have been known to go missing. This situation has developed to the point that some insurers will not provide cover for such items; even where cover is provided, in some cases (garden tools, for example) it is simply not economically viable to pursue claims — but the annoyance and expense of replacement are still there. Again, the Loop alarm will help considerably at home — it can be used to protect, a garden shed, and/or even alarm a back gate — and is priced at a fraction of the cost of a new set of 40 ft aluminium ladders!

Circuit description

Power is a nominal 6 V d.c. (though it will work with supplies of between 4 and 12 V) supplied from four AA alkaline cells. The unit draws a maximum of 70 μA when armed, giving up to 1,190 days on standby from a fresh set of batteries! The current consumption increases to approximately 56 mA when triggered. Of course, this reduces the standby life according to how long the unit had previously been left sounding — it pays to check the batteries at least twice a year! Theoretical battery life for a unit when triggered, and left to sound away, is over 35 hours.

As you can see from the circuit diagram of Figure 1.25, the circuit is built around IC1, a 4001UBE quad unbuffered NOR gate, of which two sections (b and d) are used. When power is applied, the loop is closed (the loop must present a resistance of less than 90 kΩ to the circuit, which should not pose too many problems unless your loop encircles the planet!). Under normal non-criminal circumstances, one of the inputs of IC1a (pin 1) is held low by the low resistance of the loop (pull-up resistor R4 has a value of 100 k, and thus has negligible effect). Also held low (this time by the pull-down resistor R3) are the output (pin 11) of IC1c, and the other input (pin 2) of IC1a.

Since both inputs of IC1a are low during normal operation, its output (pin 3) — and therefore the input (pin 12) of IC1c — will be high. After less than 100 ms, C1 will be charged, and IC1c (pin 13) will be held low; these components are present to ensure a power-on reset

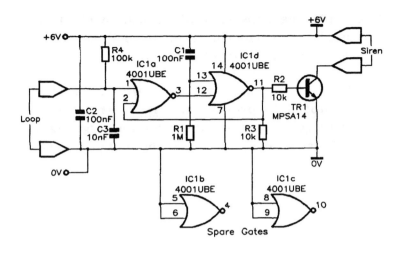

Figure 1.25   Loop alarm circuit diagram

of IC1. Since both inputs to IC1c are low, its output, and consequently the base of TR1 will also be low. TR1 is therefore switched off, and so the siren does not sound.

When the loop is *broken* — in other words, when the resistance across P3 and P4 is greater than the value of R4 i.e. 100 kΩ — the alarm condition is triggered. IC1a (pin 1) will be pulled high by R4, and so IC1a's output (pin 3) — and therefore pin 12 of IC1c — will go high. The output of IC1c (pin 11) will now go high. TR1 (MPSA14) is a Darlington device, and will conduct, therefore, applying power to the siren connected across P5 (red, +V) and P6 (black, 0 V).

Since the output of IC1c (pin 11) is also connected to the other input of IC1a (pin 2), the output of IC1a will be held low, irrespective of any change to IC1a's other input

69

Home security projects

(i.e. pin 1). Due to this innovative use of a NOR gate, the alarm is therefore latched on, and the only way to reset it is to remove power (via the keyswitch).

C3 is a decoupling capacitor, which has been incorporated to prevent any stray RF from taxi transmitters, CB sets and the like from triggering the alarm. Note that all connections to the unused gates have been grounded.

Construction

The Loop Alarm is, by virtue of its simplicity, easy to construct. To make things even easier, the PCB is designed to fit into one of the slots within a MB5 box — refer to Figure 1.26.

First, the box should be drilled — the required dimensions are given in Figure 1.27. A number of holes have to be drilled on the base and sides of the box for the siren, keyswitch, and the two phono sockets which connect to the loop itself. Once the drilling is complete, all items bar the PCB, of course, can be fitted — refer once again to Figure 1.26. The siren should be attached to its mounting bracket; the assembly can be held to the base of the box using two No. 6 x $^3/_8$ in self-tapping screws. Of course, the mouth of the siren should point towards the outlet grille drilled in the side of the box. The keyswitch and chassis mounting phono sockets can then be passed through their respective mounting holes, and secured into position with the nuts supplied. Finally, provision needs to be made for the battery holder. Self-

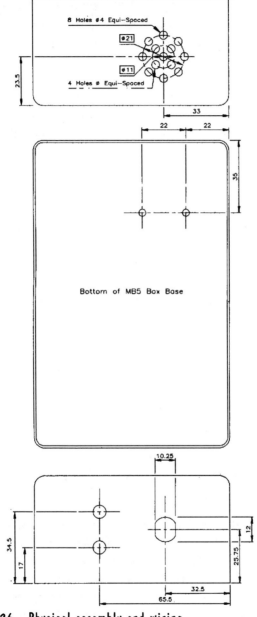

Figure 1.26   Physical assembly and wiring

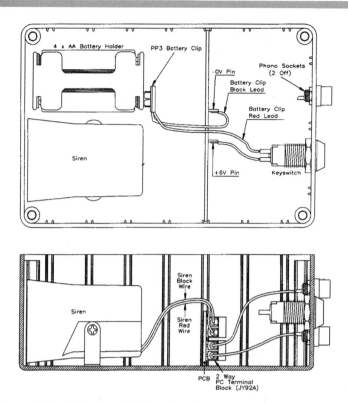

Figure 1.27 MB5 box drilling details

adhesive Velcromounts should be applied to the area of the case next to the siren (see Figure 1.26), the mating surfaces being attached to suitable positions to the 4 x AA battery holder.

Having drilled and *kitted out* the box, it's now time to put together the PCB — a simple task, thanks to the low component count. If you are not experienced in electronic construction, please refer to the Constructors' Guide supplied in the kit — a goldmine of helpful practical advice.

Referring to the component legend of Figure 1.28, insert and solder all resistors, capacitors and the two PCB pins. Next, IC1's 14-pin DIL socket — the notch in its body should line up with that of the socket's symbol on the PCB legend.

Do not fit IC1 at this stage. TR1 should follow next — please note that, like IC1, this component is polarised, and should line up with its outline on the PCB legend. Finally, the two 2-way screw terminal connectors now need to be fitted — push the connectors into position until they are flush against the board (the receptacles for the wires should point to the nearest edge of the board, to facilitate their easy insertion) and then solder them into position. PCB assembly is now complete, but it is advisable to check your work before continuing any further.

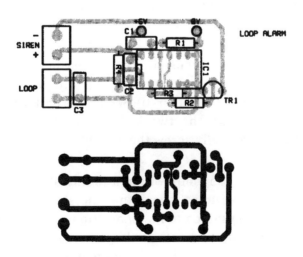

Figure 1.28   PCB legend and track

Home security projects

Now that the PCB is complete, wiring can commence; reference should again be made to Figure 1.26. Solder the black lead of the PP3 battery clip to the *0 V* PCB pin, and the red lead to one of the terminals of the keyswitch. Solder a 120 mm length of red wire (supplied in the kit) between the keyswitches other terminal, and the *+6 V* pin on the PCB.

Cut the cable attached to the siren in half, since it is far too long for this application; do not discard the excess, though, as it will be used shortly. The siren should be connected to the appropriate screw terminals — the red wire goes to the *+siren* connection. The siren's excess wire, put to one side earlier, should now be used to connect the centre terminals of the two phono sockets to the screw terminals marked *loop* (the outer terminals of the sockets are not used). After inserting IC1 (ensuring correct polarity) the board can now be fitted into the second or third slot of the box (from the keyswitch end), with the track side of the PCB facing the siren.

Fit four AA alkaline cells into the battery holder (watch the polarity!) and clip the battery clip into place (check that the key switch is off first!). Secure the battery holder in place, mating the two Velcromounted surfaces together, and finally screw the box lid on.

Having warned anybody within a radius of 100 m of an impending loud noise, check that the siren sounds by turning the keyswitch to its *on* position — if you hear a loud noise reminiscent of a wailing banshee, then switch off and continue; if no noise is apparent, you have a fault and should check your work.

Connect a suitable lead between the two phono sockets (the only requirement here is that there is a connection between the two centre conductors). One could be temporarily borrowed from your Hi-Fi system for the purpose of this test; alternatively, such a lead could be made up from a spare piece of wire and a couple of phono plugs.

With the lead in place, switch the alarm on. The alarm should not operate at this point, but as soon as the phono lead is disconnected at either end, the siren should sound. Reconnecting the lead should not have any effect — the only way to silence it is to switch it off!

Installation

Now that the unit has been tested, you can now proceed to install it. The loop, terminated with phono plugs at either end, and of suitable length must now be made, capable of reaching all of the items for which protection is intended; a suitable cable is the single-core lapped audio variety (XR13P). Cable wiring is illustrated in Figure 1.29. Since the resistance can be as high as 90 kΩ the loop can effectively be as long as you like! The alarm itself should be hidden out of view (e.g, under a desk) preferably somewhere near the cashier, supervisor or manager's working area. Of course, the unit must be installed somewhere dry — this comment is particularly valid when the unit is being used to protect items outside the home or shop. Nevertheless, the fact that the loop length is restricted by practical considerations, rather than by any limitations of the design, may be of assistance here.

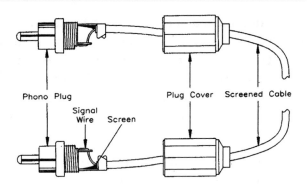

Figure 1.29   Making up the loop cables

When installing the loop, pass it through the main body of the item to be protected. For example, when protecting a bicycle, pass it through the frame and the wheels. When protecting a ladder, pass it through several rungs. Cameras and the like could be protected by passing the loop through a permanently-attached carrying strap. If the item to be protected does not have any suitable apertures through which to pass items (e.g., TV set), a suitable one could be glued on, or a more elegant situation could be devised. For example, a computer could have an unused connector, or one with at least two unused pins, and these could be used to form a loop. Again, only your imagination is the limit here. The golden rule to remember is that if a thief thinks he is going to receive hassle when trying to steal something, he is likely to reconsider and move elsewhere.

If at a later stage you need a longer loop, you could make a second lead to cover the extra distance, and connect them *in the middle* using *barrel* connectors (HH05F).

Loop alarm parts list

Resistors — All 0.6 W 1% metal film (unless specified)

R1	1 M	1	(M1M)
R2,3	10 k	2	(M10K)
R4	100 k	1	(M100K)

Capacitors

C1,2	100 nF 16 V minidisc	2	(YR75S)
C3	10 nF 50 V disc	1	(BX00A)

Semiconductors

IC1	4001UBE	1	(QL03D)
TR1	MPSA14	1	(QH60Q)

Miscellaneous

micro piezo siren	1	(JK42V)
MB5 ABS box	1	(YN40T)
min key switch	1	(FE44X)
14-pin DIL socket	1	(BL18U)
PP3 clip	1	(HF28F)
2-way 5 mm PCB terminal block	2	(JY92A)
4 AA battery box	1	(HF29G)
velcromount	1	(FE45Y)
no. 6 x 3/8 in self-tapping screw	1	(LR67X)
10 m 7/0.2 wire red	1	(BL07H)
phono chassis socket	2	(YW06G)
PCB	1	(GH46A)
instruction leaflet	1	(XU32K)
constructors' guide	1	(XH79L)

Home security projects

Optional (not in kit)

alkaline AA	4	(FK64U)
screw-cap phono black	2	(HQ54J)
single cable grey	as req	(XR13P)
phono adaptor male	as req	(HH05F)

2 Locks

Keycode lock

Are you tired of always losing your keys? Would you like a high-tech way of locking up your valuables? Perhaps the Keycode Lock is what you have been looking for. It operates in the same way you see on many a spy or sci-fi film, that is, a small numerical keypad is positioned by the door: and to open the door, the correct correct code must be entered.

It doesn't have to be confined to doors though it could be used on cupboards, filing cabinets, desks, cars, caravans, trailers, lock-ups, garages, equipment, and so on. In fact, anything that normally requires a conventional lock to secure a door or lid, and enough room to be able to fit a solenoid lock — the number of applications is infinite.

Circuit description

Switches 1 to 9 on the keypad are arranged in a matrix. A four digit code is set up by fitting links in the circuit to select a combination, which registers as Codes A to D. The complete Keycode Lock unit is shown in a block diagram in Figure 2.1.

To assist the reader whilst reading the description, refer to the circuit diagram in Figure 2.2. Code A is the first digit (1 to 9), Code B is the second, Code C and D are the third and fourth digits; all other digits must be connected to the *not used* line.

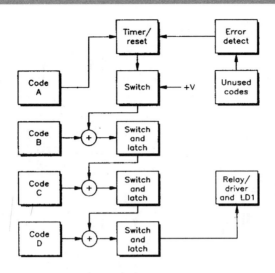

Figure 2.1 The block diagram of the keycode lock

If the first digit is entered correctly the analogue switch ES1 control line is connected to +V, causing the analogue switch contacts to become closed. Also when the first correct Code A is entered, capacitor C1 charges via resistor R9. The charge on capacitor C1 ensures that the analogue switch ES1 remains closed. Resistor R1 discharges capacitor C1 to provide the time limit for entering the four digit code (approximately 5 seconds). To inhibit this time limit fit the link J1; the capacitor then charges via the resistor R10; ES1 remains closed until an unused code digit is entered.

When an incorrect (unused) digit is entered, transistor T2 is switched on; this discharges the capacitor C1 and connects the control line of ES1 to 0 V, causing the switch to become open circuit.

81

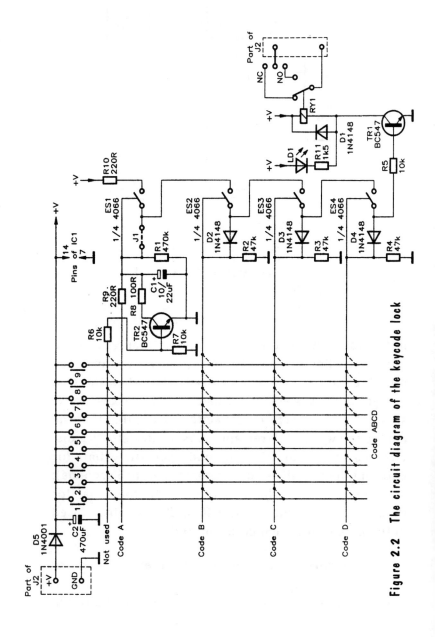

Figure 2.2 The circuit diagram of the keycode lock

Entering the second correct digit switches on ES2; the +V supply latches ES2 on via R10, ES1 and D2.

Entering the third and fourth digits correctly switches on and latch ES3 and 4 in the same way as ES2.

The analogue switches ES1–4 only latch in the order: Code A, Code B, Code C, Code D, because of the cascading arrangement of the circuit.

When all analogue switches (ES1–4) are closed, transistor T1 switches on; LED (LD1) illuminates, and the relay (RY1) operates.

The diode D1 is required to prevent damage to the driver transistor T1, from the induced e.m.f., during the decay of the magnetic field within the relay coil.

Setting the code

To set the codes, the wire links are positioned in different locations; at first this may seem a little confusing.

In deciding the codes, select any of the switches for the first digit in the code sequence, or use the same switch for all four digits if this is required.

Code A is the first pressed in the sequence, B the second, C the third, D the fourth; all other unused switches are connected to the *not used* line.

Example, if the code 1234 is required, then the following links are made as shown in Figure 2.3 (see also Table 2.1).

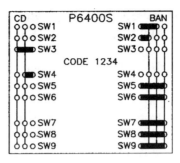

Figure 2.3 The wire links set up for code 1234

Construction

The Codelock kit comprises two PCBs: the switch PCB and the relay PCB. Build either of them first but the construction of the keypad switch PCB is described first. If you are new to project building, refer to the *Constructors' Guide* (XH79L) for details of how to recognise, handle and solder specific types of components.

Fit the light emitting diode (LED) LD1 (observing correct polarity) according to the legend on the PCB. Next fit the nine miniature tactile switches on the PCB, making sure that they are flush against the board.

Decide on what code is required for the switch, set the code with wire links, and solder in position, making sure that they are as close to the PCB as possible, so as not to touch the aluminium front panel. In Figure 2.3, the example shown is 1234. Any four numbers will do, just as long as they are remembered — pick a number, like your pet goldfish's birthday!

Switch	Code
SW1	A
SW2	B
SW3	C
SW4	D
SW5	N
SW6	N
SW7	N
SW8	N
SW9	N

Table 2.1   Example switch-code settings

Before assembly of the supplied seven bandoliered wire links, cut the individual lengths of wire as shown in Figure 2.4, and locate in position on the board. Solder the wires on the track side, making sure that the wires on the component side are trimmed to prevent accidental shorting against the aluminium front panel.

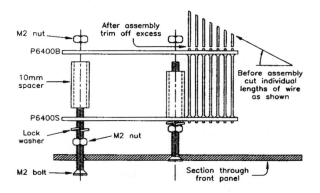

Figure 2.4   The exploded assembly including the fitting and cutting of the seven interconnecting wires

Constructing the second PCB

Locate position *J* on the PCB and with one of the bandoliered wire links, fit and solder in. If latching operation is required, as mentioned in the circuit description, then fit and solder in position J1 a small wire link. If latching operation is not required then leave the position open circuit.

Before fitting the relay, again decide on whether to fit a wire link between *NC* for normally closed contact of the relay, or fit *NO* for normally open contact use. Note that once the relay is fitted the link will *not* be accessible.

Identify the diodes, and fit and solder in, according to the legend on the PCB. Preform the leads for horizontal or vertical mounting. Note that the bar on the diode denotes the cathode and this is marked *c* on the PCB for the vertically mounted diodes and by a bar on the horizontal. Next identify the resistors, again the leads should be preformed before mounting the resistors on the board and soldering in position.

Next identify the electrolytic capacitors. The positive lead is normally longer, the negative is denoted by (–) symbols running down the body of the capacitor. Fit the capacitors with the positive lead denoted by a (+) symbol on the board, and solder in position.

Fit the 16-pin DIL IC socket, with the notch matching that on the PCB legend. Make sure before soldering in position, as it is very difficult to desolder the leads if a mistake is made.

Next fit the transistor according to the legend on the PCB and solder in position.

Locate the 4-way terminal block onto the PCB, and mount with the terminal contacts pointing away from the board, and solder the leads.

Next mount the relay; it will only go in one way, and then solder, making sure that it is located squarely over the wire link.

Finally insert the IC, with the notch on the IC correctly orientated with the notch on the legend and the IC socket.

Fitting the boards together

The boards are now mounted back to back and fitted together as shown in Figure 2.4. First pass the two M2 bolts through the aluminium panel, the heads are located into countersunk holes. Then fit an M2 nut and lock washer to each, and tighten up the M2 nuts. Now locate the switch PCB onto the M2 bolts, and make sure that the LED is in its correct position. Readjust if necessary.

Next fit the two 10 mm plastic spacers onto the M2 bolts. Carefully mount the relay PCB onto the two M2 bolts, and at the same time locate the seven wires from the switch PCB, passing them through the PCB. Fit the two remaining M2 nuts onto the bolts and ensure that the two PCBs are securely bolted together as there is no access to the screw heads once the front membrane panel is fitted. Now solder the seven wires on the track side of the relay PCB, and trim off the excess.

This completes construction of the Keycode Lock. Before attaching the membrane front panel, mount and fix the unit into the enclosure it is to operate from using security screws, available from all good hardware stores.

If using the Keycode Lock outside then it is essential to build it into a weatherproof enclosure. The membrane front panel is splash-proof only.

Using the keycode lock

Applications for the Keycode Lock are many, however, for it to operate it will need an external power supply.

External connections to the Keycode Lock are shown in Figure 2.5. The relay connections are available on the connector block J2, and would have been set earlier by the link under relay RY1 to either normally open (NO) or normally closed (NC). The power connections are also fed to J2; note the ground is marked as GND on the PCB.

To test the unit, simply turn on power to the unit, key in the code previously set, and the onboard relay RY1 will operate. Also the front panel LED will illuminate. To reset the lock, push any unused key.

A suggested wiring layout is given in Figure 6(a), using the Keycode Lock with an external power supply unit (PSU), and a door lock mechanism with an external PSU.

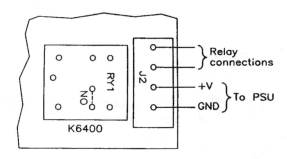

Figure 2.5   External connections to the terminal block on the PCB

A suitable PSU for the door lock mechanism can be constructed in a large PSU box (YU31J). Figure 6(b) shows the box drilling of the door lock mechanism PSU. Fit the mains transformer into the large PSU box with the fuses as shown in Figure 6(c).

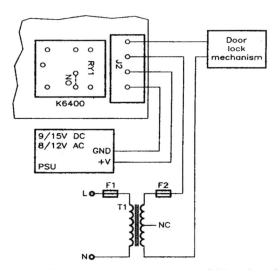

Figure 2.6(a)   Suitable interconnections to PSU and to door lock mechanism

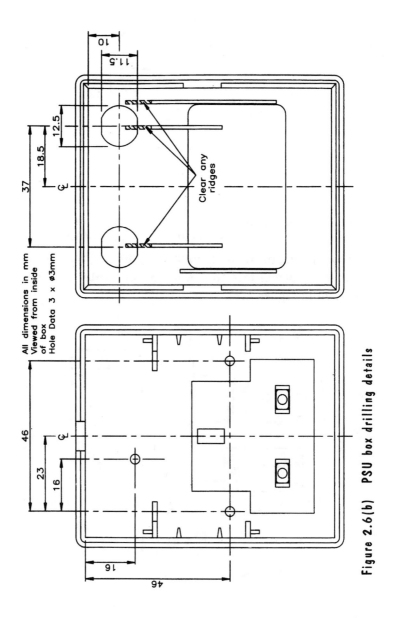

Figure 2.6(b)   PSU box drilling details

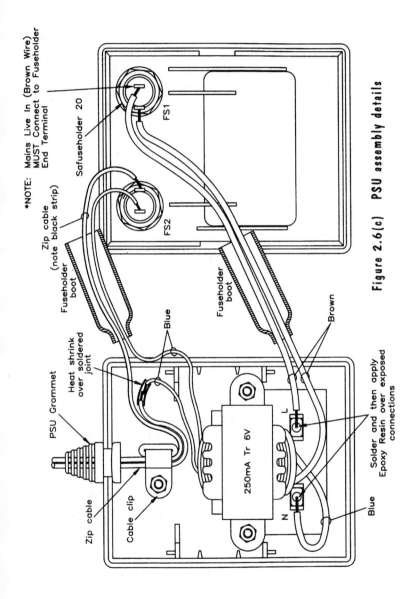

Figure 2.6(c)   PSU assembly details

*NOTE: Mains Live In (Brown Wire) MUST Connect to Fuseholder End Terminal

Safuseholder 20

FS1

Zip cable (note black strip)

Fuseholder boot

FS2

Fuseholder boot

Brown

Blue

Heat shrink over soldered joint

PSU Grommet

Zip cable

Cable clip

250mA Tr 6V

L

N

Solder and then apply Epoxy Resin over exposed connections

Blue

Don't forget to fit the fuseholder insulation boots! The transformer secondary centre tap, which is not used, cut short and then insulate with heat shrink sleeving. Fit the cable exit grommet to the output lead (note that XR39N *zip wire* is used for this purpose), connect to the transformer secondary and insulate with heat shrink sleeving. Before cutting off any excess lead, ensure that there is enough wire to reach from the nearest mains socket to the mechanism. Secure the zip wire with a *P* cable clip. Reassemble the case; a multimeter set to its a.c. range and connected across the output wires will read 13 V or so when the unit is plugged in. Once testing is complete, glue the two halves of the box together.

Switching an external device

Figure 2.7 shows the Keycode Lock operating a mains rated relay. The PCB mounted connector and track are not suitably spaced for voltages above 50 V, and if mains voltages are switched then use an off-board mains rated relay. The external relay will require a diode to protect the circuit from the induced e.m.f. produced by the relay when it de-energises.

Another application is to use an opto-isolated switch; Figure 2.8 shows the connections to the zero crossing opto-switch (LP55K).

Using the optional Mains Opto Switch kit (LP55K) with the Keycode Lock enables resistive mains loads of up to 250 W (maximum) to be switched.

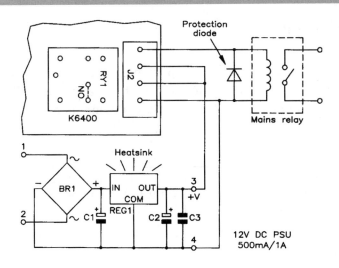

Figure 2.7   Mains relay and suitable d.c. PSU

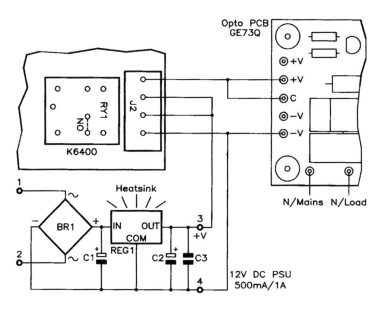

Figure 2.8   Zero crossing opto switch (LP55K)

Home security projects

Warning

It is imperative that every possible precaution is taken to prevent electric shock. Please take great care when using the LP55K module, as 230 V a.c. mains *can kill.*

Power supply	9 to 15 V d.c. or 8 to 12 V a.c.
Current consumption	Off 0.3 µA, on 40 mA
Time limit for code input	5s (only in pulse mode)
Dimensions	80 x 80 x 40 mm

Table 2.2   Specification

Keycode lock parts list

Resistors

R1	470 k	1
R2–R4	47 k	3
R5–R7	10 k	3
R8	100 Ω	1
R9,10	220 Ω	2
R11	1k5	1

Capacitors

C1	10 µF 63 V electrolytic	1
	or 22 µF electrolytic	1
C2	470 µF 25 V electrolytic	1

Semiconductors

T1,T2	BC547	2
D1–D4	1N4148	4
D5	1N4001	1
IC1	MC14066	1
LD1	3 mm LED red	1

Miscellaneous

	sub-miniature tactile switches	9
	4-way terminal block	1
RY1	5 A/220 V relay	1
	14-pin DIL IC socket	1
	metal front plate	1

Home security projects

plastic membrane	1
M2 10 mm plastic spacers	2
M2 20 mm bolt	2
M2 nut	4
wire jumpers	20
PCB P6400B1	1
PCB P6400S	1
constructors' guide	1 (XH79L)

Optional (not in kit)

solenoid lock mechanism	1 (YR88V)
mains opto switch	1 (LP55K)

The above parts (excluding optional) are available as a kit, order as VE76H

Lock mechanism PSU

PSU box large	1 (YU31J)
PSU grommet	1 (JM16S)
250 mA 6 V transformer	1 (YN14Q)
F50 mA 20 mm fuse	1 (WR93B)
T315 mA 20 mm fuse	1 (RA07H)
20 safuseholder	2 (RX96E)
fuseholder boot	2 (FT35Q)
$^3/_{16}$in cable P clip	1 (LR44X)
zip wire	1 (XR39N)
2-core 3 A mains cable	1 (XR47B)

+12 V 500 mA PSU

C1	1000 μF 63 V axial electrolytic	1	(FB84F)
C2	10 μF 63 V radial electrolytic	1	(JL10L)

C3	100 nF disc ceramic	1 (BX03D)
BR1	W01	1 (QL29G)
REG1	L78M12CV	1 (QL29G)
	500 mA 9 V miniature	
	transformer	1(WB11M)
	0.5 A/1 A PSU PCB	1 (YQ40T)
	heatsink	1 (HQ81C)
	1.3 mm PCB pin	1 (FL21X)
	6BA $^1/_2$in bolt	1 (BF06G)
	6BA nut	1 (BF18U)
	6BA washer	1 (BF22Y)

Infra-red doorlock

Have you ever been in the following situation? Your hands are full of shopping, it's dark and the porch light is off. The kids are playing you up and, during the struggle to get the key into the door, you drop the weekly groceries, breaking the eggs and glass jars. If this wasn't bad enough, milk from the split containers causes the labels to peel off from the tins, and ruins anything else that may have been salvageable, and in a burst of frustration and anger, you shout at the kids.

Well, this project will solve all (sorry, *nearly* all) of your problems. It will unlock the door, and can turn on the (porch or hall) light for a few minutes (with the addition of a relay, or the LP55K Mains Opto Switch Kit) by a simple press of a button.

This project has many other applications — opening garage doors, for example.

The infra-red transmitter

The infra-red transmitter, shown in Figure 2.9, outputs an amplitude-modulated carrier. As you can see from the circuit diagram in Figure 2.10, this carrier is generated by op-amp IC2 (LF351) and associated circuitry.

IC1 (M145026) provides a stream of data corresponding to the conditions set on pins 1 to (high, low, or floating). This data is used to modulate the carrier and is transmitted from LD2, via TR1.

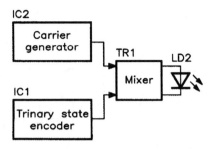

Figure 2.9 Intra-red transmitter block diagram

The receiver

The receiver is shown in block form in Figure 2.11, and Figure 2.12 shows the circuit diagram. Information received by the infrared diode, IR1, is processed by the receiver IC1 (TDA3047), which contains an HF amplifier with a 66 dB control range, a synchronous demodulator, a reference amplifier, an AGC detector, a pulse shaper, an input voltage limiter and a Q-factor killer. It strips off the carrier, leaving only the bare digital data stream.

This data stream is then passed to IC2, the companion to the remote control encoder in the transmitter, which will only respond (via pin 11, which becomes active) if the information contains the correct address, as set in the transmitter. There are 6,561 different codes possible, only one of which will trigger the device — security is therefore assured.

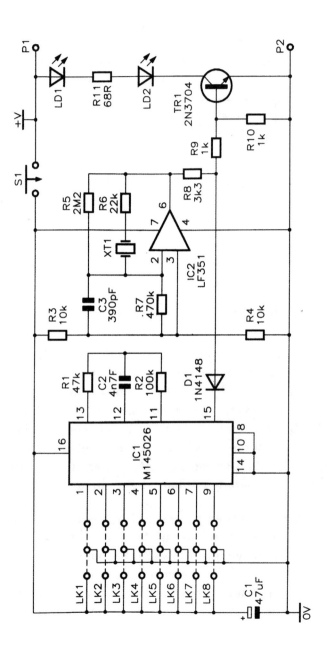

Figure 2.10 Intra-red transmitter circuit diagram

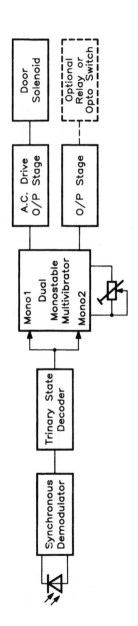

Figure 2.11   Infra-red receiver block diagram

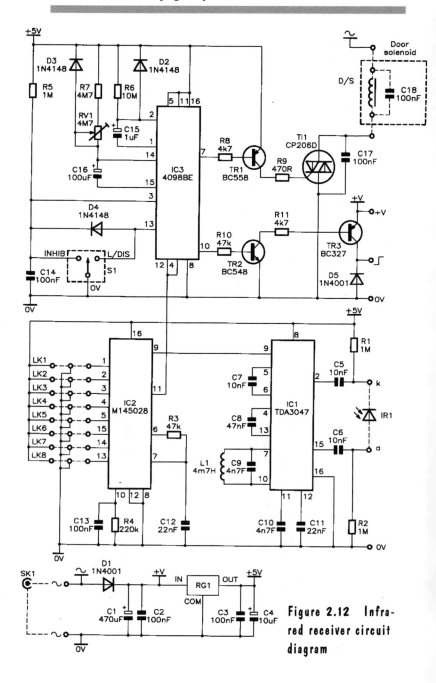

Figure 2.12 Infra-red receiver circuit diagram

IC3 is a dual monostable; once triggered by IC2, one half of the dual monostable will operate the door lock, via TR1 and TI1, for 3 seconds (determined by the time constant of R6 and C15) — enough time to open the door. The other half of IC3 has a variable timed output (determined by R7, RV1 and C16) ranging from 1 to 2 minutes. TR2 and TR3 are used as a voltage level shifter, which would be useful to turn on a porch or hallway light — but not directly; the LP55K Mains Opto Switch Kit or a relay should be used as an *interface* — more details on this later.

Constructing the infra-red transmitter

A fibreglass PCB has been chosen for maximum reliability and stability. To allow the PCB to fit into the key-ring remote case, no IC sockets have been used, so please double-check orientation of the ICs before fitting, as removal after incorrect insertion will almost certainly damage both the PCB and the chip! For further information on component identification and soldering techniques, please refer to the *Constructors' Guide* included with the kit.

Figure 2.13 shows the PCB, with printed legend, to help you correctly locate each item. The sequence in which the components are fitted is not critical; however, the following instructions will be of use in making these tasks as straightforward as possible.

Start by fitting resistors R1 to R11, taking note that R5 is a larger, metal film resistor. Using offcuts from the leads

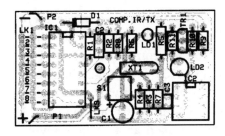

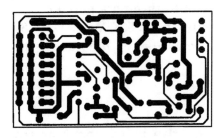

Figure 2.13   Transmitter PCB legend and track

of these resistors, insert link LK9. LK1 to LK8 set up the code transmitted. With reference to Figure 2.14, there are three ways of connecting the links (well, two ways, the third is to leave them out!). The purpose of these codes is to make your transmitter *unique*, so that only you will be able to open your door.

Insert S1 and XT1, bending the crystal flush to the PCB as per the legend.

Insert and solder C1 to C3, taking care of the orientation of C1. The polarity of this capacitor is shown by a plus sign (+), matching that on the PCB legend. However, on the actual body of most electrolytic capacitors the polarity is designated by a negative symbol (−), in which case the lead nearest to this symbol goes in the hole nearest the edge of the PCB.

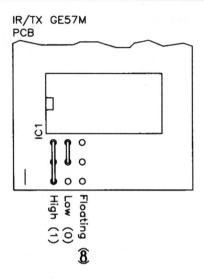

IR/TX GE57M
PCB

IC1

High (1)
Low (0)
Floating (8)

Figure 2.14   Three ways of fitting transmitter links LK1–LK8

With reference to Figure 2.15, insert LD1 at a height of
7 mm. The best way to do this is to cut a thin strip of
card 7 mm wide, and place this between the legs of LD1
and the PCB whilst it is soldered. Bend the legs of LD2 at
90° as shown in Figure 2.16, and insert into the PCB at a
height of 3 mm.

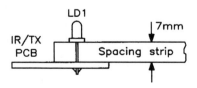

Figure 2.15   Inserting LD1 at a height of 7 mm

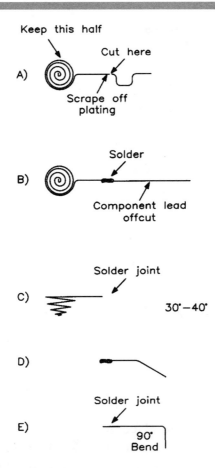

Figure 2.16   Battery terminal spring bending

Identify the two battery clips. The positive terminal (a small metal plate) is soldered into the slot labelled + or *P1*. The negative terminal, or *spring*, is prepared as shown in Figure 2.17, before being soldered in the hole marked − or *P2*. The spring may need bending slightly to allow it to fit snugly inside the box.

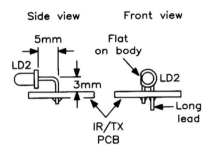

Figure 2.17   Fitting LD2

Insert diode D1, taking care of its orientation. The band at one end corresponds to the thick white line on the legend. TR1 is inserted matching its package outline with that of the legend. Install the ICs, making sure that all the pins go into the holes, and that the pin number one marker or notch at one end of the IC package matches up with the white block on the legend. Take *great* care while soldering the ICs in place, allowing several seconds between solder joints for the IC to cool down! As a guide, do the four corner pins first, ensuring that the IC is flush with the board. You can solder the remaining pins at your leisure, with long pauses in between.

Boxing up the infra-red transmitter

If the remote key-ring box comes with two switch actuators fitted, then these need to be removed by pulling/breaking the actuator shaft. File the lower of the two inserts out to the dimensions shown in Figure 2.18. A

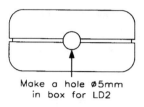

Make a hole ⌀5mm
in box for LD2

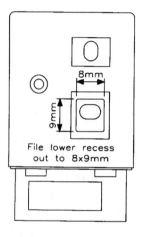

Figure 2.18   Filing dimensions for transmitter box

5 mm hole also needs to be cut in the top end of the box for LD2, as shown. This is best achieved by carefully filing each half of the box with a round file, making frequent checks by placing the PCB inside the box to ensure that a correctly sized cut-out is achieved.

With reference to Figure 2.19, cut the flexible membrane into two pieces, 7.5 mm and 10.5 mm long. Remove the paper backing and stick them into the upper (small) and lower (large) recesses respectively.

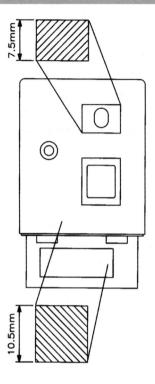

Figure 2.19   Fitting the membrane to the transmitter box

Double-check your PCB to make sure there are no dry joints or short circuits, and assemble the box around the PCB, inserting the battery (not supplied in the kit) taking care of its polarity (marked on the bottom of the battery compartment). The whole assembly is screwed together, with the single screw provided.

The only thing left to do now is to depress the larger membrane, upon which the red LED, LD1, should flash very quickly.

Home security projects

The receiver module

There are at least two ways of installing the receiver, one of which is looked at in detail here. For flexibility, the IR sensor can be mounted on or off-board; as a result, the sensor can be mounted remotely if required. The wiring diagram in Figure 2.20 is common to all configurations, however.

The receiver PCB could be housed within a double pattress fitted into (or onto) a wall, and the IR receiver diode (IR1) located remotely — in a window, perhaps. Depending on your requirements, a surface-mounting or a flush-mounting pattress could be used — each has its advantages. A double pattress provides sufficient room for the receiver module and the power supply transformer. If the Mains Opto Switch is to be used to switch a porch light, for example, this could be mounted, together with the mains transformer, in the double pattress — the receiver module could then be mounted in a single pattress. This has the safety advantage of keeping the high-voltage mains supply isolated from the low-voltage electronics. *Please note* that if any items working at mains voltages (i.e. mains transformer, Mains Opto Switch) are installed in a metal case (e.g., flush-mounting pattress), the case *must* be earthed!

If desired, the unit could be mounted in a box (ABS box MB2 is ideal), with the controls on the front and the receiver diode on the rear — in this case IR1 is mounted directly on the PCB. The box can then be placed on a window sill, with IR1 facing towards the glass and the world outside. If this option is to be followed, the power

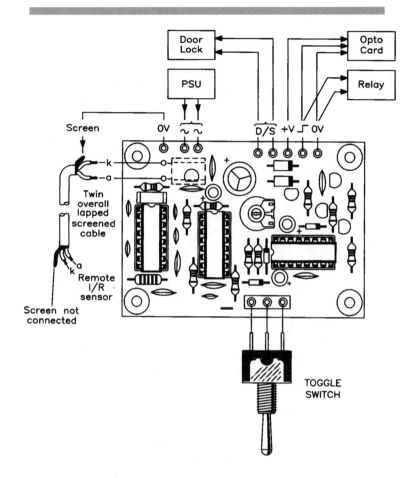

Figure 2.20   Receiver wiring

supply transformer should be mounted in a separate enclosure (for example, that shown in Figure 2.21).

Another option is to house the receiver in a single pattress, and the power supply transformer built into a separate enclosure. This latter option is featured here.

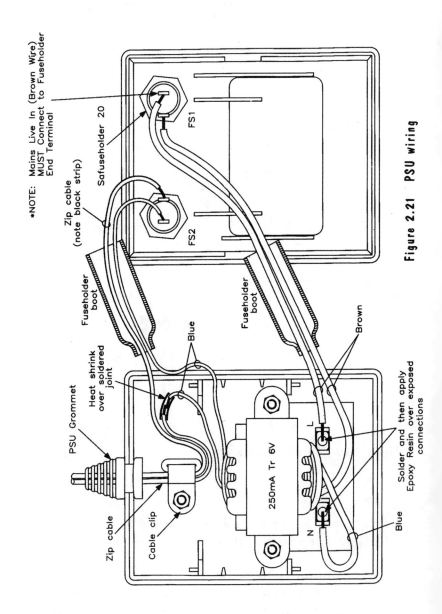

*NOTE: Mains Live In (Brown Wire) MUST Connect to Fuseholder End Terminal

Safuseholder 20

Zip cable (note black strip)

FS1

FS2

Fuseholder boot

Fuseholder boot

Brown

PSU Grommet

Heat shrink over soldered joint

Blue

Zip cable

Cable clip

250mA Tr 6V

Solder and then apply Epoxy Resin over exposed connections

Blue

Figure 2.21 PSU wiring

Constructing the receiver PCB

Construction is fairly straightforward, and inexperienced readers are directed to the *Constructors' Guide* supplied with the kit. The PCB legend and track layouts are reproduced for your convenience, in Figure 2.22. First, fit the resistors and inductor — after soldering the components in position, trim off the excess leads; these can now be used for the link. Fit the diodes; D1 and D5 are 1N4001 power diodes, while D2, D3 and D4 are 1N4148 signal diodes, which are smaller. In each case, polarity is important, and the band on the diode should be lined up with that shown on the PCB legend. It is also important to align the three IC sockets correctly — these are all 16-pin types. In each case, the notch should be aligned with the corresponding one on the PCB legend.

We can now proceed with the capacitors, C1, C4, C15 and C16 are all electrolytic types — it is essential to insert these the correct way round. The negative symbol embossed on each capacitor must face away from the + symbol on the PCB legend. The other capacitors are non-polarised and can be fitted either way round. The transistors (TR1 to 3), thyristor (TI1) and regulator (RG1) can now be fitted; their outlines should correspond with those on the PCB legend — it is important to fit these devices the correct way round.

The next step depends on how you wish to use the Infrared Door Lock. If the IR diode (IR1) is to be fitted remotely, two PCB pins should be soldered in the *a* (anode) and *k* (cathode) positions, to take the interconnecting wires; note that they should be inserted

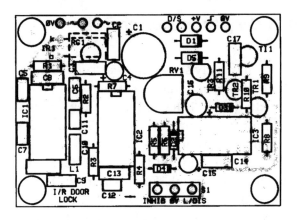

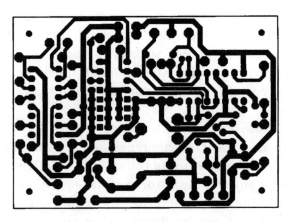

Figure 2.22   Receiver PCB legend and track

from the *track* side. However, if IR1 is to be mounted on the PCB itself, it should be fitted on the *track* side of the board. Once the IR diode arrangements have been taken care of, the other PCB pins can be fitted (from the *track* side).

Referring to Figure 2.23, the code may now be set; it should, of course, correspond to the code to which the transmitter has already been set! Finally, thoroughly inspect your work and, when you are finally happy, insert the three ICs into the sockets. The module is now ready to be installed into the appropriate enclosure.

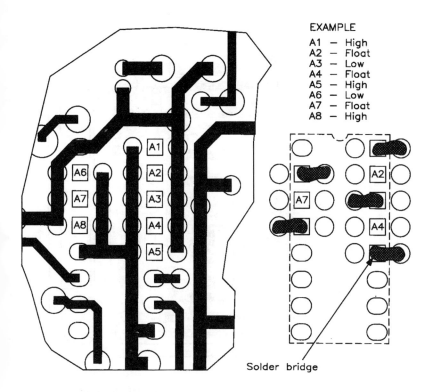

Figure 2.23   Three ways of fitting receiver links LK1–LK8

Final construction and installation

Decide where you are going to install the receiver pattress, and prepare the mounting surface (e.g. drill and install wall plugs). Obviously, it is advisable to choose somewhere near a power socket, the door and an appropriate window. Mount the receiver into a single pattress, using epoxy glue to hold it in place, as shown in Figure 2.24.

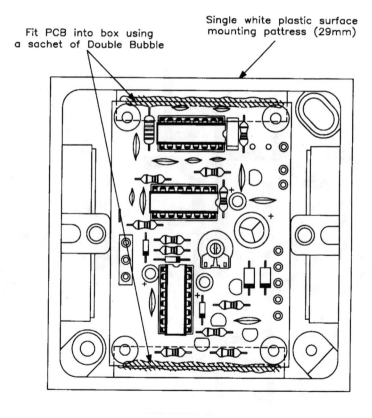

Figure 2.24 Fitting the receiver PCB to a surface-mounting single pattress

Note that a surface-mounted pattress is used for ease of construction; if a metal flush-mounted type is used, then the board needs to be mounted on spacers to avoid the possibility of short-circuits.

Next, we move onto the receiver power supply, which is shown in Figure 2.25. Fit the mains transformer into a large PSU box with the fuses, as shown in Figure 2.21. Drilling details are given in Figure 2.26. Don't forget to fit the fuseholder insulation boots! The transformer secondary centre tap, which is not used, should be cut short and insulated with heat shrink sleeving. Fit the cable exit grommet to the output lead (note that XR39N *zip wire* is used for this purpose), connect to the transformer secondary and insulate with heat shrink sleeving. Before cutting off any excess lead, ensure that there is enough wire to reach from the nearest mains socket to the pattress. Secure the zip wire with a *P* clip. Re-assemble the case; a multimeter set to its a.c. range and connected across the output wires should read 13 V or so, when the unit is plugged in.

Remove the cover plate from the door catch release mechanism and wire a 100 nF capacitor across the solenoid's two screw terminals, as shown in Figure 2.27. At this stage, the lead-in wires can also be connected; again,

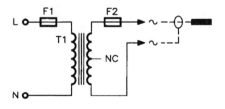

Figure 2.25   PSU circuit diagram

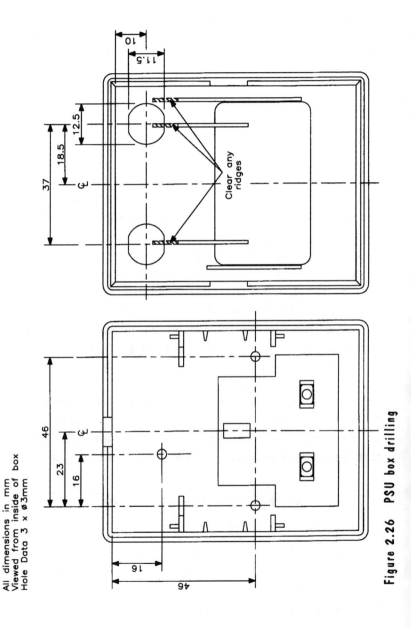

All dimensions in mm
Viewed from inside of box
Hole Data 3 × ⌀3mm

Figure 2.26 PSU box drilling

zip wire is used. After the cover plate has been replaced, the mechanism can be fitted to your chosen door. Safely route the lead to where the receiver will be located, and cut off any excess.

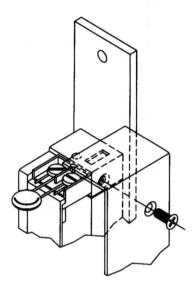

Figure 2.27   Door solenoid wiring

Break through one of the cable inlets of the pattress — this will accept the power, solenoid and IR diode wires. Connect a length of twin overall lapped screen cable (XR20W) to the points *a, k* and *0 V* (screen to *0 V*) as shown in Figure 2.20, and feed it through the cable inlet hole in the pattress. Solder the IR diode to the cable's other end — remember that polarity is important — and insulate with heat shrink sleeving; the screen of the cable is trimmed short. After passing the power supply's output lead through the pattress hole, solder each wire to one of the two – (a.c.) terminals — polarity is not important here. The final connection is the one from the door solenoid. Pass the lead through the cable inlet hole, and solder each wire to one of the *D/S* pins. Again, polarity is not critical.

Install the pattress in the previously prepared position, and fit the IR diode so that it faces out of the appropriate window; alternatively, fit a *spy hole*, and mount the IR diode in it.

Drill a 7 mm hole in the pattress blanking plate, which will accept the toggle switch. Before you drill the hole, it must be noted that the switch must be located in a position such that it does not foul the PCB. Wire the switch (mounted on the blank pattress) to the *INHIB, 0 V* (centre/common) and *L/DIS* pins of the module.

Testing the system

Fit the pattress blanking plate, and set the toggle switch to the centre position. Apply power to the system. Step outside with the IR transmitter (and your door keys, just in case it doesn't work first time!). Point the transmitter at the receiver diode, and press the button — the door lock should release for approximately 3 seconds. Close the front door, press the transmitter button again and push the door when you hear the buzzing sound of the door lock — the door should open. Please note that any pressure exerted on the lock will prevent it from releasing, Repeat the test with the toggle switch in the other two positions, one of which should inhibit the lock. The other position will inhibit the external switching circuit (i.e. the relay or Mains Opto Switch). With the switch in the centre position, the system operates normally (door and external switch outputs active).

Switching an external device

Please note that all work must be carried out with the mains power removed!

Using the optional Mains Opto Switch kit (LP55K) with the Infra-red Door Lock enables a resistive mains load of 250 W (max.) to be switched at the same time as the door solenoid. A porch or hall light would be ideal for use with such a system. The actual *on* period is longer than that of the door solenoid — and can be varied between 1 and 2 minutes using RV1.

If you are building a system from *scratch*, it is recommended that you build the system into two pattresses — a double one for the transformer and Mains Opto Switch, and a single one for the receiver. Remember that if metal pattresses are used, the PCBs must be mounted on pillars, and that the pattresses must be earthed. Since the power supply is permanently installed, there is no need for a fuse on the transformer's secondary winding (in the PSU unit in the *standard* version, this is present to protect the transformer from a short-circuit trailing lead). If you are expanding the system already described, you can fit the Opto Switch in another single pattress located next to the existing one in which the receiver module is located.

Home security projects

Once you have decided where the Mains Opto Switch is to go, complete the wiring as shown in Figure 2.28. The switch across the *N/Load* and *N/Mains* terminals of the Mains Opto Switch is a bypass switch, so that the porch light may be switched on irrespective of the state of the door lock. Of course, all connections at mains potential must be treated with the appropriate care and attention.

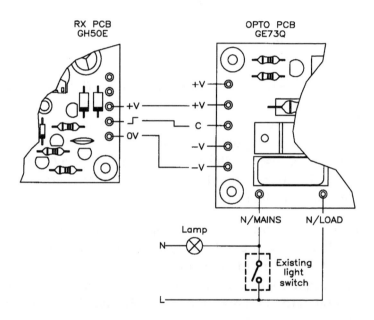

Figure 2.28 Using the mains opto switch to control a porch light or similar load

As an alternative to the Mains Opto Switch, a relay may be used to switch an external load. This option should be used if the load to be switched exceeds 250 W (a 500 W halogen lamp, for example), but note that a relay with suitably-rated contacts is used.

Warning

It is imperative that every possible precaution is taken
to prevent electric shock. 240 V a.c. mains *can kill.* Any
modifications to fixed premises wiring must be carried
out in accordance with BS7671 Requirements for Electrical Installations — IEE Wiring Regulations 16th
Editions. If in doubt as to the correct way to proceed,
seek advice of a suitably qualified person before continuing.

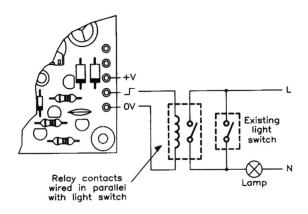

**Figure 2.29   Using a relay to control a porch light or similar
load**

I/R door lock parts list

Resistors — All 1% metal film (unless specified)

R1,2,5	1 M	2	(M1M)
R3,10	47 k	2	(M47K)
R4	220 k	1	(M220K)
R6	10 M	1	(M10M)
R7	4M7	1	(M4M7)
R8,11	4k7	2	(M4K7)
R9	470 Ω	1	(M470R)
RV1	4M7 hor enclosed preset	1	(UH11M)

Capacitors

C1	470 µF 35 V PC electrolytic	1	(FF16S)
C2,3,13, 15,17,18	100 nF 50 V disc	6	(BX03D)
C4	10 µF 50 V PC electrolytic	1	(FF04E)
C5,6,7	10 nF ceramic	3	(WX77J)
C8	47 nF polylayer	1	(WW37S)
C9,10	4n7F ceramic	2	(WX76H)
C11,12	22 nF ceramic	2	(WX78K)
C15	1 µ.F 100 V PC electrolytic	1	(FF01B)
C16	100 µF 10 V PC electrolytic	1	(FF10L)

Semiconductors

IC1	TDA3047	1	(UL25C)
IC2	M145028	1	(UJ51F)
IC3	4098BE	1	(QX29G)
TR1	BC558	1	(QQ17T)

TR2	BC548	1 (QB73Q)
TR3	BC327	1 (QB66W)
D1,5	1N4001	2 (QL73Q)
D2,3,4	1N4148	3 (QL80B)
RG1	µA78L05AWC	1 (QL26D)
TI1	CP206D	1 (UR25C)
IR1	infra-red photodiode	1 (YH71N)

Miscellaneous

L1	4.7 mH choke	1 (UK80B)
	electric door lock	1 (YU89W)
	pin 2145	1 (FL24B)
	16-pin DIL socket	3 (BL19V)
	PCB	1 (GH50E)
	compuguard IRT	1 (LP24B)
	instruction leaflet	1 (XU24B)
	constructors' guide	1 (XH79L)

Optional (not in kit)

	MB2 ABS box	1 (LH21X)
	PSU box large	1 (YU31J)
	PSU grommet	1 (JM16S)
T1	250 mA 6 V transformer	1 (YN14Q)
F1	50 mA 20 mm fuse	1 (WR93B)
F2	315 mA A/S fuse	1 (RA07H)
	20 safuseholder	2 (RX96E)
	fuseholder boot	2 (FT35Q)
	filter red	1 (FR34M)
	2.5 mm standard power plug	1 (HH62S)
	2.5 mm panel mount power skt	1 (JK10L)
	$^3/_{16}$ in cable P clip	1 (LR44X)

Home security projects

M3 10 mm steel screw	1	(JY22Y)
M3 steel nut	1	(JD61R)
M3 steel washer	1	(JD76H)
M3 isoshake	1	(BF44X)
zip wire,	1	(XR39N)
lapped pair	1	(XR20W)
twin mains DS black	1	(XR47B)
47 mm double surface pattress	1	(YB17T)
double blank plate	1	(ZB49D)
mains opto switch	1	(LP55K)
5 A mains relay	1	(YX98G)
M3 10 mm insulated spacer	1	(FS36P)
F31 SR grommet	1	(LR47B)
sub min toggle B	1	(FH01B)
3202 wire brown	1	(XR34M)
3.2 mm spade	1	(JH64U)
CP32 heat shrink	1	(BF88V)
double bubble sachet	1	(FL45Y)

The above parts (excluding optional) are available as a kit, order as LT32K

Infra-red code-lock

In any security system, the quality of the locking device determines the level of difficulty presented to an intruder attempting to gain access. Ordinary mechanical key or combination locks are vulnerable as direct physical access to them is often possible. This makes picking the lock or trying different combination codes a distinct possibility. However, with an electronic remotely controlled code-lock, an additional level of security is provided since no direct access to the locking device is necessary. In such a system the transmitter key generates an electronic code, which is then picked up and decoded by the receiver lock.

A very sophisticated integrated circuit (IC), capable of handling a total of 60,000 possible codes, has been used in the design of the system. The method of transferring the code from key to lock could use any of the following systems; a.c. induction, ultrasonics, radio frequencies, or infra-red. In this particular kit, infra-red was chosen as it offers good immunity from many of the unwanted external influences.

The transmitter

As can be seen from the circuit diagram reproduced in Figure 2.30, the transmitter appears to be deceptively simple in its design. This is because the majority of electronic functions are handled within IC1, a MM57410N

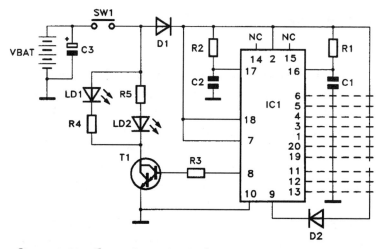

Figure 2.30 Transmitter circuit diagram

code generator/decoder. It is equipped with twenty terminal pins for connections to the rest of the circuit, although not all of these are used in this particular application. The positive power supply is fed to pin 2, with the negative connection on pin 10. Pin 18 is used to place the IC into its code generating (high), or decoding (low) modes. Ten pins (1, 3, 4, 5, 6, 11, 12, 13, 19, 20) are reserved for setting the code value. Each of these can be placed in one of three different conditions, greatly increasing the total number of possible code combinations. The code is fed out on pin 8 of the IC as a serial data stream. To ensure that the entire code is sent each time power is applied, a reset pulse is generated by R2 and C2 on pin 17. The rate at which the code is transmitted is determined by the timing components, R1 and C1, on pin 16.

To drive the red LED, LD1, and the infra-red transmitter, LD2, transistor T1 is used as a saturated switch. It is turned on when pin 8 of IC1 is high, thus modulating the red and infra-red LEDs with the serial code data.

Transmitter construction

Included in the transmitter kit is an instruction booklet which lists the complete step-by-step construction details. The surface mount devices (SMDs) are already bonded to the transmitter PCB, leaving only seven additional standard components to be fitted to the board. However, care must be taken not to damage any of the SMDs when soldering or cutting off the excess component leads. The step-by-step building instructions are easy to follow, and construction should proceed smoothly. The only items that require mechanical preparation are two small metal strips, which form the battery holder.

Transmitter testing

Before testing can commence you must first carefully install four LR44 type button cells, which are not included in the kit (Maplin Order Code FM28F). It is of great importance that you are very careful in observing the correct polarity of each cell in the battery and that the total polarity matches the PCB markings. It might be necessary to bend the battery strips a little, in order to obtain sufficient pressure on the cells.

For the initial tests, the default code is sufficient until you require to change it to your personalised code — see *Defining the code*. The default, or standard, code condition exists while all the code connections of IC1 are in the *open* position (not soldered to an adjoining pad). At this stage, it is not possible to perform any conclusive infra-red transmission tests; to do this would require a known working receiver module. However, since the red LED, LD1, is modulated by the generated code it should light up every time that SW1, the push switch, is pressed. Once in the keyholder box, the circuit board and the cells are held securely in place by the battery cover.

The receiver

As can be seen from Figure 2.31, the infra-red receiver circuit has a more complex structure, incorporating several integrated circuits. The infra-red signals from the transmitter are detected and converted back into electronic signals by the infra-red detector diode D1. However, these signals are at a very low level and require boosting if the code-lock system is to work successfully over a reasonable distance. This is achieved by IC1 which contains four operational amplifiers, shown in Figure 2.31 as A1 to A4.

The final amplified signal is taken from the output of A3 and is connected to the *code input* (pin 7) of IC2, another MM57410N. In its receive mode, this IC compares the data signal applied here to the code value set on its programming pins (1, 3, 4, 5, 6, 11, 12, 13, 19, 20). When

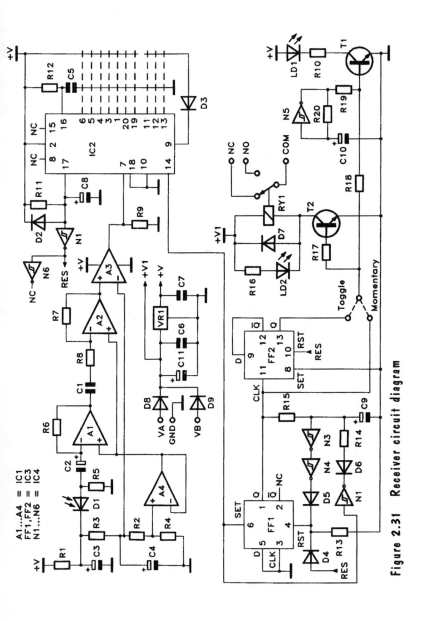

Figure 2.31 Receiver circuit diagram

131

a positive code match has been found, the logic state of pin 14 is flipped over. The rest of the receiver circuit is concerned with processing this change of logic state so that it can switch the relay and indicators.

Both momentary and toggle switch actions are available from the outputs of the dual D-type flip flop FF1 and FF2 (both contained within IC3). A *high* logic state, from the mode selected, is used to forward-bias transistors T1 and T2. T1, the receive mode indicator switch, is normally turned on and off at a regular rate controlled by N5, a Schmitt inverter contained within IC4. This causes LD1, a red LED, to flash, indicating that the receiver is on and ready to receive an infra-red code transmission. In the momentary switch mode this indicator will be permanently lit while a valid code is being received. However, when the toggle mode is selected and a valid code received, it will stay permanently lit until the code is re-transmitted.

The relay, RY1, and T2, the transistor switch that controls indicator LD2, are normally turned off. In the *momentary* mode, RY1 will be activated and LD2 will light while a valid code is being received.

When the *toggle* mode is selected and a valid code is received, both will stay permanently active until the code is re-transmitted.

Receiver construction

Included in the receiver kit is an instruction booklet which lists the complete step-by-step construction details. The receiver circuit is split into two PCBs; a small pre-amplifier board and a larger main decoder PCB. Once built and checked over, the pre-amp module is placed inside a metal screening box. This is done to reduce the possibility of stray electrical noise swamping the infra-red receiver and causing unwanted switching problems. Two PCB configurations are described and, depending on the physical constraints of your intended installation, one of them should prove more applicable than the other. The first method can be used where space is no problem, and involves mounting the metal boxed pre-amp module directly onto the main decoder PCB, as shown in the photograph of the receiver. The second method should be used where not enough room is available to accept both PCBs. In this case, the small pre-amp module is separated from the main board. No specific case is supplied, as its design will vary with your requirements. The clear plastic cases supplied with the transmitter and receiver will do at a pinch, though. If you opt for this approach you could paint them in the colour of your choice, leaving a clear unpainted area for the view of the infra-red receiving device.

Receiver testing

To test the functions of the receiver you will require a suitable power supply, a multimeter and an infra-red code-lock transmitter. Before commencing the tests you must set the operating mode link to either the momentary or the toggle position. For the initial testing the default code is sufficient. This default, or standard, code condition exists while all the code connections of IC2 are in the *open* position (i.e. not soldered to an adjoining pad). This code pattern must be matched to that transmitter if the lock is to function.

Either low voltage a.c. or d.c. can be used to power the unit, and the connections are made using screw terminals on the PCB. If you want to run the unit off 230 V a.c. mains supply, you must use a step-down transformer with a 9–0–9 V centre-tapped secondary winding capable of providing 100 mA. The Maplin sub-miniature type (Order Code WB01B) is well suited for this task. The secondary of this transformer should be connected to the PCB terminals as follows; connect the two end wires to VA and VB, and the centre wire to GND. If a d.c. power supply is to be used, it should be capable of providing 100 mA at +10 to +16 V. This voltage should be applied to VA or VB, the negative connection being made to the GND terminal.

The current consumption is approximately 20 mA with LD1 flashing in the receive mode, increasing to 77 mA when activated by the infra-red code transmitter. These figures were obtained when running the unit off a +12 V regulated power supply. Check that both LD1 and LD2 are lit and that the relay RY1 clicks. Finally, test the range of the codelock system, which under ideal conditions should be approximately ten metres.

Defining the code

Up until now, the default code has been used during the testing of the code-lock system. However, this code value is only one out of the possible 60,000 that can be set by the user. In other words, even if someone else has a similar infra-red transmitter key, that person still only has a one in 60,000 chance of having the same code.

On both the transmitter and receiver PCBs, the code ICs have ten programming pins (1, 3, 4, 5, 6, 11, 12, 13, 19, 20). Each pin can be placed in one of three possible conditions:

●   not connected to any other solder pad,

●   soldered to the pad positioned on the outside of the IC pin,

●   soldered to the pad positioned on the inside of the IC pin.

The exact positioning of these pads are clearly shown in the supplied instruction booklets. You can make the code pattern as symmetrical or as random as you like, but do check that the patterns match on both receiver and transmitter PCBs.

Installation

Before installing the code-lock system, you should first determine whether its power requirements can be supplied from an existing source, or if a separate a.c./d.c. power supply is necessary. Another important consideration is the switching action of the relay and its electrical characteristics. For safety reasons, it is strongly recommended that the 240 V a.c. mains is *not* applied directly to the relay terminals on the receiver board. In most vehicle and building applications, the infra-red receiver will probably be located behind glass. The type of glass used can possess quite different degrees of transparency at infra-red wavelengths, and this will ultimately determine the distance over which the system will work. The range can also be adversely affected by strong sunlight, or other infra-red sources.

Applications

Although not infinite, the possible number of applications is just too great to cover in a single chapter. As a result, only four of the more popular uses have been highlighted.

Probably the most popular use of the infra-red code-lock is as a remote arm/disarm switch in vehicle and home security systems.

Another very popular application is as an electric door catch controller, and Maplin can supply a solenoid-activated mortise lock release mechanism (Order Code YU89W).

The system could possibly be used in conjunction with a garage door opener, enabling you to gain access as you approach the doors.

General remote power switching of a.c. and d.c. currents not exceeding 10 A, and at voltages less than 50 V, is relatively simple and safe to use for lighting or motor control.

136

Transmitter	Receiver
60,000 possible code combinations	60,000 possible code combinations
Transmit range 10 metres	Receive range 10 metres
Housed in key-ring box	Output mode: toggle or momentary
SMD components pre-mounted	Relay output: 10 A at 12 V d.c.
LED indication of transmission	LED display for receive and output status
Power supply: 4LR44 button cells	Power supply: 2 x 9 V a.c. or 10 to 16 V d.c.
	Supply current: 100 mA

Table 2.3   Technical specifications

Transmitter parts list

Resistors

R1,2,3, 4,5	all surface-mounted devices (SMDs)	5

Capacitors

C1,2	SMDs	2
C3	47 µF 16 V electrolytic	1

Semiconductors

1C1	MM54710	1
TR1	Darlington SMD	1
D1,2	1N4148	2
LD1	subminiature LED red	1
LD2	CQW15 LED infra-red	1

Miscellaneous

PCB	1
case, with battery cover	1
sub-miniature tactile push-to-make switch	1
material for constructing battery terminals	1 sheet
construction/user guide	1

Optional (not in kit)

LR44 button cell	4 (FM28F)

The above parts (excluding optional) are available as a kit, order as VE10L

Receiver parts list

Resistors

R1,2	47 Ω	2
R3,4,17, 18,19	10 k	5
R5	100 k	1
R6,7,20	1 M	3
R8	33 k	1
R9,14	470 Ω	2
R10	270 Ω	1
R11,12, 13	47 k	3
R15	470 k	1
R16	680 Ω	1

Capacitors

C1	4n7 ceramic	1
C2,8,9, 10	1 µF 25 V electrolytic	4
C3,4	22 µF 25 V electrolytic	2
C5	33 pF ceramic	1
C6,C7	100 nF monolithic ceramic	2
C11	470 µF 16 V electrolytic	1

Semiconductors

LD1,2	rectangular LED red	2
D1	BPW41 I/R receiver diode	1

Home security projects

D2,3,4,		
5,6,7	1N4148	6
D8,9	1N4000 series rectifier diode	2
TR1,2	BC547 (or equivalent)	2
IC1	LM324 quad op-amp	
	(or equivalent)	1
IC2	MM57410	1
IC3	4013	1
IC4	40106	1
VR1	7806 voltage regulator	1

Miscellaneous

	14-pin DIL socket	3
	20-pin DIL socket	1
	PCB pins	8
	6-way screw connector	1
RY1	single pole changeover relay	1
	pre-amp PCB	1
	main PCB	1
	screening components for	
	pre-amp PCB	2
	construction/user guide	1

Optional (not in kit)

case (presentation box
supplied in kit could be used —
see text)
power supply components

The above parts (excluding optional) are available as a kit, order as VE09K

3 Miscellaneous

Featuring:

Infra-red proximity detector

Commercially available body-heat-movement-detection systems, although very sophisticated in their operation, can be rather expensive for use in limited applications where short-range coverage is required. This I/R proximity detector has been designed as a simple, low-cost system for detecting heat changes, movement of a warm body, etc, such as those emitted from the human body. The unit responds to a definate change or disturbance in ambient — or background — heat levels and could be placed across a doorway or stairs to register movement in those areas.

Pyroelectrics

The E100SV1 sensor uses a ceramic, ferroelectric element, which has the property of producing an electrical change due to a change in polarisation intensity. If a moving object enters the field of view of this sensor, changes in infra-red energy levels occur due to a difference in temperature between this object and the background. Infra-red energy is converted into heat by the surface electrode of the element, thus causing a change in temperature within the element itself, and a small electric charge is created as a result (see Figure 3.1).

This small charge appears across the gate resistance Rg in Figure 3.2, and is impedance buffered by the FET source follower, where a change in voltage appears across source resistance R13 (Figure 3.4).

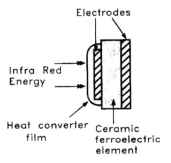

Figure 3.1   Pyroelectric element

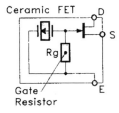

Figure 3.2   Proximity detector — internal

A small d.c. bias voltage is produced across R13 by the quiescent current flowing through the FET while no signal is present, as can be seen in Figure 3.3. Output signals from the source terminal modulate the d.c. bias level with a $+V_e$ voltage swing.

In use, the voltage swing is very small, its amplitude being determined by the amount of incident energy available, which becomes smaller with increasing distance.

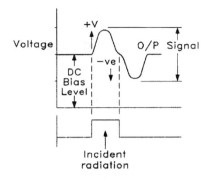

Figure 3.3 Source output voltage swing

Done with mirrors!

A negligible amount of energy is emitted from the human body which limits the effective working range of the module down to four feet or so. This range could be extended by increasing the sensitivity of the amplifier and developing velocity related filter circuits which would determine a given range of movement speeds and size of body.

An even more effective method is employed on commercial systems in the form of collecting lenses and optical amplifying concave mirrors. Problems associated with energy collecting systems are: movements in the air, sunlight *modulated* through curtains and even small animals generating fluctuations in the infra-red energy background. To help overcome these sorts of problems, a multi-faceted, concave mirror is often used, which has the effect of expanding (or narrowing) the field of view into bands.

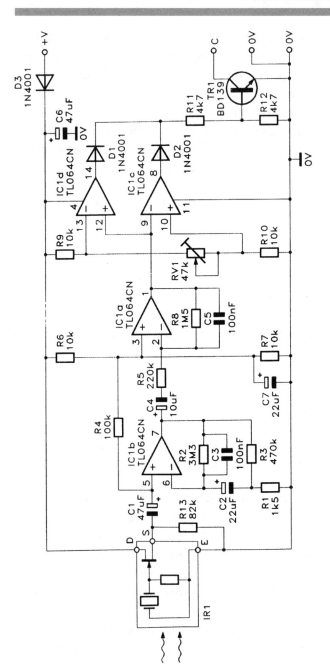

Figure 3.4 Circuit diagram

Home security projects

As an infra-red emitting source crosses the field of view, radiated energy bounces off these facets in a sequence. The sensor responds with a series of related output pulses; detection electronics can determine the size, velocity and direction of the source while it is moving. Sophisticated PIRs are readily available in the Maplin catalogue and are more suitable for security and alarm use than this particular system.

However, many applications exist where a simpler system is called for, especially for the home constructor!

Circuit description

The circuit shown in Figure 3.4 consists of two amplifying stages, with low-pass filtering and a comparator threshold stage. Output voltage swings from the PIR are amplified by IC1b, which is configured as a non-inverting amplifier. The PIR receives energy from many sources, a mixed waveform would be produced at IC1b output, therefore C3 integrates continuous low-level signals and acts as a low-pass filter.

The somewhat unusual arrangement of resistors R1 and R4 allow capacitor C2 to charge slowly during initial power up. Capacitor C2 is necessary to isolate IC1b $-V_e$ input from the 0 V supply rail. With single supply op-amps, it is common to generate a half supply d.c. voltage reference to bias the differential inputs, thus allowing output voltage swings about this level. The effect of integration on the continuous input signals produces a very low-frequency output signal, which is applied to C2.

146

The charge across C2 varies with the magnitude of the output signal (from pin 7) and limits heavy transients which might saturate this stage.

IC1a is a standard inverting amplifier, again voltage referenced to half supply by R6 and R7. C7 decouples the reference voltage to prevent comparator supply spikes from being introduced into the stage. IC1d and IC1c serve as a simple comparator. The threshold voltage reference (determining when the comparators will trigger) is set by RV1 in the potential divider chain R9 and R10.

Positive voltage swings from IC1a will trigger the IC1d comparator causing D1 to conduct, while negative swings trigger IC1c causing D2 to conduct. From Figure 3.3 it can be seen that the output voltage swing from the PIR is, firstly, in a positive direction and then secondly in a negative direction. The ultimate effect from the comparator output at R11 is, therefore, not one but two pulses turning on transistor TR1.

Either one of diodes D1 or D2 could be removed for single pulse output, and which particular one to remove must be decided under full operational conditions. TR1 is an open collector switch, which will sink external loads (sourced from their own external +V supply) to the 0 V common rail when conducting.

Construction

For information on building details and components, refer to Figure 3.5 for the board layout and to the *Con-*

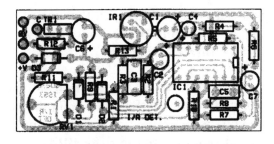

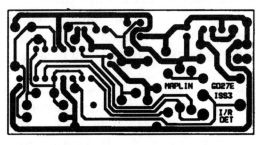

Figure 3.5 PCB legend and track

structors' Guide supplied with this kit (if you do not intend to purchase the complete kit then see the Parts List for the order code of the *Constructors' Guide*). Identify and insert resistors R1 to R13. Solder these components and remove excess wire before continuing.

Mount diodes D1 to D3, and insert the four printed circuit pins into the holes marked with white circles. Next, insert a 14-pin IC socket in position IC1 (taking note of the orientation notch), bend a few of the legs over the track pads to hold it in position. The PCB is quite small with tracks running close together, so a little extra care must be taken whilst soldering, as short circuits between tracks can easily occur.

Identify and insert capacitors C1 to C7. Polylayer type C3 should be fitted carefully to avoid breaking off the lead-out wires from the ends of the package. Fit preset RV1 and solder all components in position. Again cut off all excess leads, then fit TR1 and the sensor IR1 shown in Figure 3.6. One side of TR1 has a metal, heat transfer mounting plate fitted. Insert TR1 with this plate facing outward towards the edge of the PCB. The sensor IR1 shown in Figure 3.7 could be mounted vertically above the PCB, or horizontally off the PCB as detailed. In either case, mount the sensor as close as possible to the PCB in order to reduce noise induced into this area.

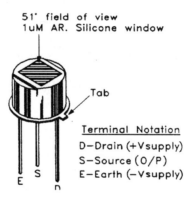

51° field of view
1uM AR. Silicone window

Tab

Terminal Notation
D—Drain (+Vsupply)
S—Source (O/P)
E—Earth (—Vsupply)

E   S   n

Figure 3.6   Sensor pin-outs

Either sensor mounting position will have to take into account the enclosure (case) requirements, and this is left to be fitted as required by the constructor. Last of all, fit a TL064 into the IC socket — taking note of the orientation; cut off all excess wires and clean up the track area to facilitate inspection.

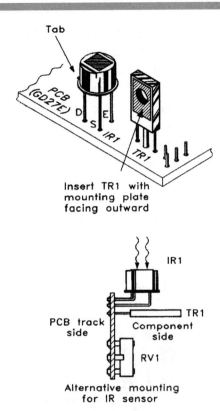

Figure 3.7 Mounting arrangements

Testing

Supply requirements for the module are 9 V d.c. at 2 to 3 mA. Current consumption is low, which allows long periods of use from small battery packs such as a PP3-sized battery. Connect the battery +V_e to the pin marked +V, and –V_e to either 0 V pin; diode D3 prevents damage to components in the event of accidental battery or supply reversal.

Check the supply current with a milliammeter which will be around 2.5 mA for a minute or so dropping to between 1 and 1.5 mA after this period. Current consumption increases by approximately 1 mA while the comparator stages are operating.

The output transistor TR1 does not source current, but being open collector will sink current from an externally supplied load. Figure 3.8 suggests various methods of switching external loads, and diagram (a) could be used

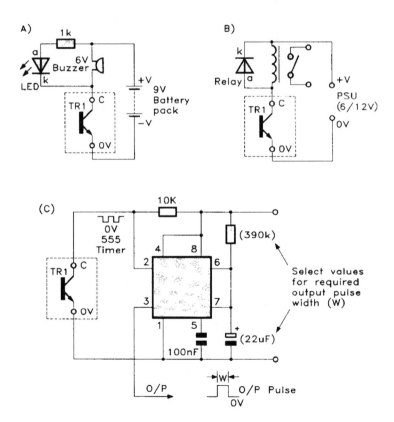

Figure 3.8 (a), (b), (c): external circuit connections

Home security projects

for testing purposes. Connect the LED cathode (k) to collector of TR1, the pin marked C, then connect a 1 kΩ resistor between the LED anode (a) and the $+V_e$ supply (battery $+V_e$ or +V pin).

If using the same battery for both module supply and LED supply, then the second battery $-V_e$ connection is not required (pin 3).

Turn the comparator threshold control RV1 to half travel (Figure 3.9) then, after the initial *warming up* period, move your hand across the sensor window. Do not poke the window with fingers as grease deposits will reduce sensitivity and may prevent operation completely! Figure 3.10 shows the spectral response expected in the window. The LED will light for a few seconds. If the LED is permanently aglow, turn the trigger level down by moving RV1 wiper anti-clockwise.

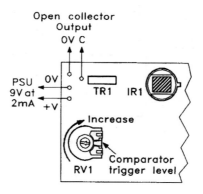

Figure 3.9   External PCB connections

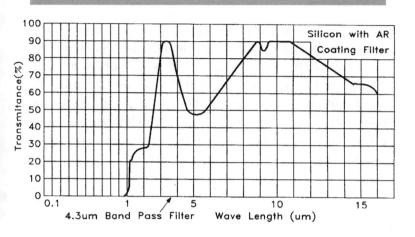

Figure 3.10   Spectral response of the sensor window

Using the module

Transistor TR1 is not capable of switching heavy loads and should be used only with external systems up to 12 V d.c. and current levels below 100 mA. Relays could be used for controlling larger voltage/current devices (Figure 3.8(b)), or a timer could be employed to generate long operating periods once triggered (Figure 3.8(c)). On the prototype, a 6 V at 35 mA buzzer was used on a separate supply to good effect.

Any battery supplying the electronics should *not* also be used for supplying the external devices as well (if more than a simple LED arrangement is to be used). Battery connections to pins 1 and 2 should be kept short — a PP3-sized battery clip lead is ideal for this — mount both module and battery together in the same housing with a suitable *on/off* switch.

Home security projects

Sensing range is 4 to 5 feet, depending upon the sensor's field of view and variations in the light/heat background levels. A whole room, for instance, could not adequately be covered by this system, but doorways, narrow hallways and corridors are suitable areas. Another use for the module could be in a shower cubicle, using a timer circuit for controlling the water pump. Obviously, low voltage switching systems are important in this application.

Infra-red proximity detector parts list

Resistors — All 0.6 W 1% metal film (unless specified)

R1	1k5 min res	1	(M1K5)
R2	3M3 min res	1	(M3M3)
R3	470 k min res	1	(M470K)
R4	100 k min res	1	(M100K)
R5	220 k min res	1	(M220K)
R6,7, 9,10	10 k min res	4	(M10K)
R8	1M5 min res	1	(M1M5)
R11,12	4k7 min res	2	(M4K7)
R13	82 k min res	1	(M82K)
RV1	47 k hor encl preset	1	(UH05F)

Capacitors

C1,6	47 µF 16 V minelect	2	(YY37S)
C2,7	22 µF 16 V minelect	2	(YY36P)
C3	100 nF polylayer	1	(WW41U)
C4	10 µF 16 V minelect	1	(YY34M)
C5	100 nF 16 V minidisc	1	(YR75S)

Semiconductors

D1,2,3	1N4001	3	(QL73Q)
TR1	BD139	1	(QF07H)
IC1	TL064CN	1	(RA66W)

Home security projects

Miscellaneous

IRD1	E100SV1 I/R detector	1 (UR69A)
	PCB	1 (GD27E)
	pin 2145	1 (FL24B)
	14-pin DIL socket	1 (BL18U)
	instruction leaflet	1 (XT72P)
	constructors' guide	1 (XH79L)

The above parts are available as a kit, order as
LT00A

Electronic watchdog

The basic idea behind the Electronic Watchdog is that while an electronic siren will deter most burglars, some still persist with the break-in. However, a furiously barking dog should make him or her think twice about continuing the attempt, and hopefully scare the villain off. For a variety of reasons it is not always possible or desirable to keep a real dog in the home, and since only its bark is required, it can be replaced by some electronics! This concept dates back to the early days of sound recording using gramophone records and tape recorders. Although quite effective, these mechanical systems suffered from long term reliability, i.e. record, stylus, tape, tape heads and motor wear. Only in recent years, with the introduction of digital solid state technology, has it been possible to significantly improve this situation, see Figure 3.11.

How do you get a dog inside an integrated circuit?

First find your dog and make it bark. Then, using a microphone to pick-up the sound, convert the analogue signal into a digital data format. Finally, store this in a nonvolatile memory IC. Sounds difficult? Don't worry, the kit's manufacturer Velleman has already done all this for you, and this specially programmed device is included in the kit (IC4, VLK2655). It has sufficient storage capacity for two different dog barks which are selected using links J1 or J2 on the printed circuit board (PCB).

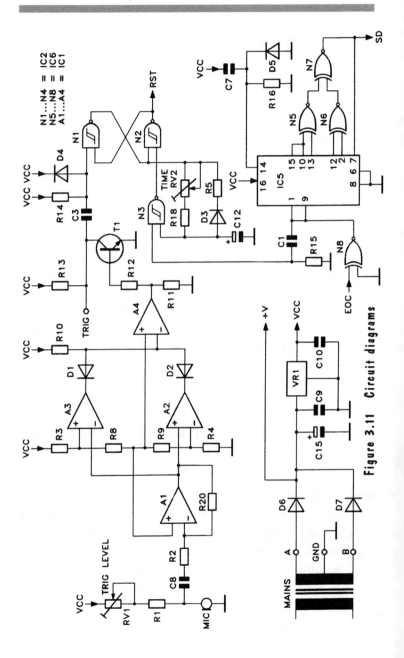

Figure 3.11 Circuit diagrams

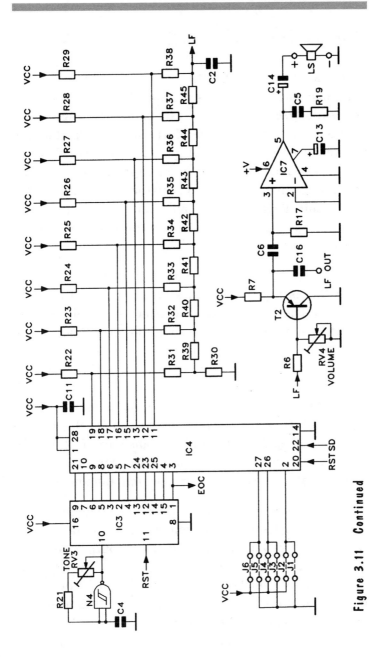

Figure 3.11 Continued

159

How do you get the dog out of the IC?

To replay the dog barks, the Read Only Memory (ROM), IC4, must have each of its memory locations accessed sequentially. This is achieved by using a 12-stage ripple counter, IC3, the count rate of which is governed by the frequency of the clock signal generated by N4 of IC2. This frequency is set by the value of capacitor C4, and the combined values of the two resistors R21 and RV3. On the circuit diagram, Velleman refers to RV3 as a *tone control*; a more accurate description would be *replay speed control.*

As IC4 steps through its memory locations, the digital recording is reproduced as a stream of eight-bit data patterns. These have to be converted back into the original analogue waveform before audio amplification can take place. This digital to analogue conversion is achieved by using a network of resistors R22 to R45, with the final audio product (LF) appearing across C2. This signal then passes through a potential divider comprising R6 and RV4, where RV4 is used to adjust the volume of the barking. An audio buffer transistor, TR2, is provided which allows both the on-board amplifier IC7 and an external amplifier to be driven simultaneously.

As the on-board amplifier is capable of producing quite a loud bark on its own when connected to a suitable loudspeaker, Velleman has omitted the LF output coupling capacitor C16 from their kit. Should you wish to fit one, a 100 nF polyester layer capacitor (Maplin stock code WW41U) will make the audio signal available at the LF output terminals. In the Velleman instruction booklet it is stated, "If you are sure that the *external* power amplifier already has a capacitor at the input, then you may fit a wire link for C16."

Don't do this! For the small additional cost of an extra capacitor, you ensure that the amplifier you intend using, or any amplifier you may use in the future, be isolated from the d.c. component of TR2.

What causes the dog to bark?

The Electronic Watchdog, just like a real dog, responds to acoustic stimulus, though of course no electronic device could hope to achieve the same degree of sensitivity over such an extreme frequency range as a real dog can perceive. However, by employing the use of a highly sensitive electret microphone insert, Velleman have made their unit very responsive to environmental noise. The degree of microphone sensitivity is set by the trigger level control RV1.

To activate the digital circuits, the audio signal produced by the microphone must be amplified and converted into a trigger pulse. This is achieved by the amplifiers A1 to A4 (IC1), and the resulting d.c. trigger pulses are fed to TR1 which acts as a transistor switch. This allows the connection of an extra external trigger pulse, which can be produced from a wide variety of electronic sensors, i.e. infra-red motion, smoke, heat, pressure mat, vibration, or a complete security system. The output of TR1 is connected to the digital circuits via C3, and this produces a narrow pulse which initiates the digital sequence of events. For each trigger pulse the *dog* will bark for a predetermined length of time before returning to its surveillance mode. The bark duration is set by RV2, and can be varied from approximately 4 to 37 seconds. If you want the dog to keep barking for as long as the external trigger is grounded (active), replace C3 with a wire link.

The kit

The Velleman kit is contained in a substantial plastic packing box and includes all the electronic components necessary to construct the finished PCB assembly. To assist in its construction, three separate instruction booklets are provided; these being English and continental building information and multi-lingual component identification. Not included in the kit though is the mains power supply transformer or loudspeaker. This is because of the differing requirements that may be placed upon the basic unit in any given situation — more about this later.

PCB assembly

The assembly instructions provided with the kit are brief, and assume a certain amount of constructional knowledge on the part of the kit builder. If you require additional information about soldering and assembly techniques, they can be found in Maplin's own *Constructors' Guide*, stock code XH79L. Removal of a misplaced component will be fairly difficult, so *please* double-check each component type, value, and its polarity where appropriate, before soldering! The PCB has a legend to help you correctly position each item; see the back page of the Velleman English building booklet.

Power supply

Owing to its relatively high standby current (75 mA maximum), the Electronic Watchdog is not suited to long-term battery operation. The two recommended supply methods are from 230 V a.c. mains supply transformer, or a 9 to 12 V unregulated d.c. power unit. The transformer specified by Velleman has an 8–0–8 V secondary at 400 mA; Maplin cannot supply a transformer with this precise specification but recommends 6–0–6 V at 500 mA (stock code WB06G). When using mains transformers you *must* follow all the relevant safety precautions to avoid electric shock! For this reason it is much safer and more convenient to use a ready made d.c. power supply, such as the unregulated 300 mA type (stock code XX09K). Although only rated at 300 mA, it nevertheless is more than capable of supplying the quoted 400 mA maximum peak current consumed by the Watchdog when barking. Both sets of power supply connections are shown in the wiring diagram, see Figure 3.12.

Loudspeaker

The range of loudspeakers to choose from is vast, but whichever you choose, it only has to match, or exceed, the following specification:

2 watts, 4 or 8 ohms.

Home security projects

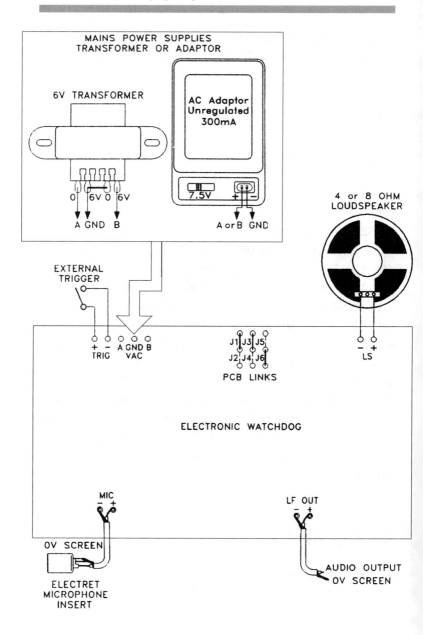

Figure 3.12 Wiring diagram

The Maplin catalogue has a large section dedicated to loudspeakers, and several are suitable for the Watchdog project, here are just a few:

3 watt	elliptical	GL16
6 watt	elliptical	GL17T
3 watt	low-cost	YJ16S
6 watt	low-cost	GL12N
5 watt	mylar	YN01B
10 watt	mylar	YN02C

To improve the acoustic performance of any of the above speakers, they must be housed in a suitable cabinet. For convenience and durability, the 10 watt horn speaker (stock code XQ73Q) offers high sound output without the need to be placed in an enclosure.

Wiring

The only wire included in the kit consists of two short lengths of tinned copper wire used for the PCB links (*jumpers* J1, J2, J3, and J6). The amount and type of additional off-board wiring is determined by your particular installation requirements. Once again, the Maplin catalogue has a section which covers a large range of cables and wires. Here are just a few that could be of use in this project:

● 3-core 3 A Mains cable black XR02B (sold per metre),

● zip connecting cable XR39N (sold per metre and ideal for loudspeakers),

Home security projects

- single-core lapped screen cable black XN12N (sold per metre and ideal for general audio connections),

- hook-up wire (7/0.2) black BL00A (sold per 10 metre pack for general interconnections).

A wiring diagram showing all the connections is given in Figure 3.12. When using mains cable, follow all the normal safety precautions to avoid electric shock.

Testing

The test procedure given in the Velleman instructions is very basic and does not involve use of any test equipment. However, if you do decide to take some instrument readings you might find a significant difference between your results and those given in the Velleman booklet! This is because the type of power supply, loudspeaker and test meter used will influence the measurements you obtain. In addition, the individual specification and tolerance of the components used in the kit will have a combined effect on the results. The following test results were obtained using a +12 V d.c. power supply, 4 ohm speaker load and a digital multimeter:

Velleman technical data	Maplin test results
Supply current:	
standby, 75 mA	16 mA
barking, 400 mA max	270 mA max

Table 3.1

Velleman electronic watchdog parts list

Resistors

R1	330 Ω	1
R2–5	10 k	4
R6	200 k	1
R7–9	1 k	3
R10–13	4k7	4
R14–17	47 k	4
R18	220 k	1
R19	10 Ω	1
R20	2M2	1
R21	56 k	1
R22–29	2k7	8
R30–38	100 k	9
R39–45	51 k	7
RV1	4k7 (or 5 k) preset	1
RV2	2M2 (or 2M5) preset	1
RV3	22 k (or 25 k) preset	1
RV4	470 k (or 500 k) preset	1

Capacitors

C1	100 pF ceramic	1
C2,3	10 nF ceramic	2
C4	3n9F ceramic	1
C5,6	47 nF ceramic	2
C7,8	100 nF polylayer	2
C9–11	100 nF monores	3
C12,13	10μF electrolytic	2
C14	220 μF electrolytic	1
C15	1000 μF electrolytic	1

Home security projects

Semiconductors

D1–5	1N914 (or 1N4148) diode	5
D6–7	1N4000 series diodes	2
TR1	BC547 (or BC548, BC549)	1
TR2	BC557 (or BC558, BC559)	1
IC1	324	1
IC2	4093	1
IC3	4040	1
IC4	VLK2655	1
IC5	4015	1
IC6	4077	1
IC7	386	1

Miscellaneous

8-pin DIL socket	1
14-pin DIL socket	3
16-pin DIL socket	2
28-pin DIL socket	1
PCB pins (8 used)	10
wire links	3
PCB	1

A kit of the above parts is available, order as VE85G

Forged bank note detector

Recently you may have noticed a strange blue glow being emitted from behind cashiers' desks and tills? No, it's not aliens taking over. It's the notorious bank-note checker, where an ultraviolet fluorescent tube is used to illuminate the bank-notes. The reason is that fake or counterfeit notes reflect the UV light whereas real ones do not.

Our idea, a slight variation on the standard type of checker, is to incorporate an extra circuit which sounds a buzzer and illuminates an LED if the note is OK, but if the note is fake then the LED will not light and the buzzer will not sound. Since the unit will operate from a +12 V d.c. supply, it is ideal for use at Car Boot Sales, where counterfeit notes are often tendered. As *Forged Bank-Note Detector* is a bit of a mouthful it will be referred to simply as the *Till Saver*.

Circuit description

The design of the circuit is very simple and comprises of two main sections; these being the tube driver and the LDR sensing circuit. The full block diagram is given in Figure 3.13. The tube driver circuit, already constructed on a small PCB, is a very simple circuit using a specially wound transformer. The detection circuit comprises of a light dependent resistor (LDR) sensor and associated components which operates a buzzer and LED.

Home security projects

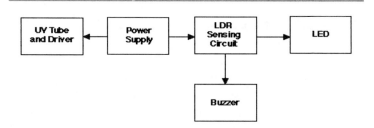

Figure 3.13   Block diagram of the Till Saver

The fluorescent tube driver circuit is shown in Figure 3.14. The power supply, (smoothed by a 47 µF capacitor) on being applied charges up the 0.1 µF capacitor via the feedback winding of the transformer and the 20 Ω resistor. The voltage on reaching approximately 0.6 to 0.7 V the transistor switches on. The 15 nF capacitor, charged via the 620 Ω, resistor then dumps its charge through the transformer primary; (the capacitor is

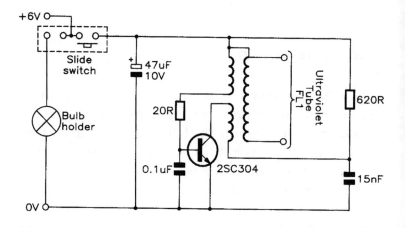

Figure 3.14   Circuit diagram of the fluorescent tube driver

170

required for the current pulse). A sufficiently high voltage develops on the secondary of the transformer to excite the gas in the UV tube and cause illumination. The feedback winding on the transformer will apply a negative voltage to the base of the transistor, the base capacitor will turn off the transistor and allow the cycle to continue. The actual oscillator frequency is approximately 60 kHZ.

Warning: the tube voltage is around 1000 V peak-to-peak so be extremely careful not to touch anything in this area when the unit is powered up.

The detection circuit is shown in Figure 3.15. The Till Saver is polarity protected by diode D1 and uses a LM317T regulator along with resistors R1 and R2 to obtain an output voltage of +6 V. Capacitor C1 provides high frequency decoupling at the input of the regulator, and capacitor C2 promotes stability at the output of the regulator. Capacitor C3 is a reservoir capacitor for the power supply to the tube driver.

The light dependent resistor (LDR), R3 and VR1 form a potential divider network, where VR1 is the sensitivity control. Under normal conditions the UV light falling on the LDR will lower its resistance and bias TR1 on; with TR2 biased off LED LD1 and buzzer BZ1 will not operate. When the amount of UV light falling on the LDR is low, its resistance will be high thus, TR1 switches off, and this in turn allows TR2 to switch on, operating the LED LD1 and sounding the buzzer BZ1. When the UV light level increases once again this will be detected by the LDR and consequently the circuitry will switch the LED and buzzer off.

Home security projects

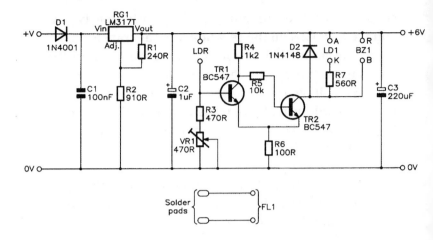

Figure 3.15   Circuit diagram of the Till Saver

Construction

Referring to the PCB legend and track in Figure 3.16 and the Parts List, start by fitting the resisters and the vertical preset resister. Then fit the diodes, making sure that they are orientated correctly, with the band at the cathode end matching that on the PCB legend.

Next fit the capacitors, noting that the electrolytic and tantalum types are polarised. The electrolytic capacitor negative lead, which is shown by a black band and (−) symbols is fitted towards R6 and the longer lead (+) towards R4 and R5. Fit the tantalum type with the (+) marked on the body towards the (+) symbol on the PCB.

Next fit the transistors, making sure that the outlines match those as shown on the PCB legend.

172

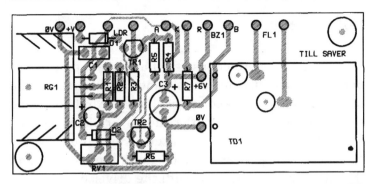

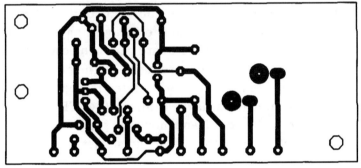

Figure 3.16 PCB legend and track

The PCB pins should be fitted next, and can be pushed home with the tip of a soldering iron and soldered in position.

Next bend the legs of the regulator to enable it to lay flat on top of the heatsink, and place the legs in the appropriate holes. Bolt the regulator and heatsink into place using an M3 x 10 mm bolt, nut and shakeproof washer; then solder the regulator legs.

Sub-assembly

To complete the project use the fluorescent tube driver board from an Ultraviolet Mini Lantern. This needs to be taken out of the case and can be done by carefully taking the casing apart. Remove the end cover plates and the clear plastic cover and then the ultraviolet tube. A white plastic covering needs to be taken off the red case and the case can be split apart with the driver board easily extracted. The two white wires can be cut or desoldered. Cut the curly battery terminals off leaving an excess of about 8 mm. The tube driver board can then be mounted with the cut down terminals soldered to the pins on the main PCB marked 0 V and +6 V, (refer to Figure 3.17), these are located towards the centre of the PCB. Remove the lamp from the board, and move the slide switch fully in the direction away from the lamp socket.

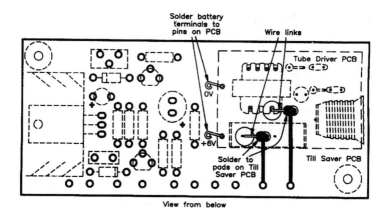

Figure 3.17   Mounting the Fluorescent Driver PCB onto the Till Saver PCB

Two links must be made from the fluorescent tube driver output, and passed through the holes in the main PCB, to the pads on the board leading to the pins marked FL1. Connect the LDR to the pins marked on the PCB, bending the leads so that the LDR is set at 90° to the board. Connect the LED (The Cathode (K) denoted by the short lead and the flat side of the package) to the pins marked A and K. Connect the buzzer with the black wire to pin B, and the red wire to pin R. Connect the power lead from pins 0 V and +V to the 2.5 mm power socket, making sure that the +V lead is connected to the centre pin.

The design of the module allows it to be mounted on to a piece of 39 x 62 hole matrix board (Order Code JP53H). The drilling details of the matrix board are shown in Figure 3.18; along with an exploded view showing the mounting of the two PCBs onto the matrix board which should then be completed using two 10 mm insulated spacers (FS36P). Also fit the two fuse clips in position which will enable the fluorescent tube to be placed onto the board. The PCB assembly can then be housed in an enclosure such as ABS box H2852 (BZ75S).

Final assembly

Figure 3.19 shows the box drilling and cutting details. Drill the holes as shown, including a hole at the front for the LED and at the rear to allow for the adjustment of VR1.

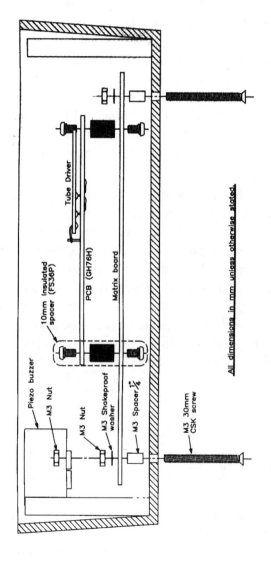

Figure 3.18 Matrix board drilling details and exploded view showing mounting of the PCB onto the matrix board

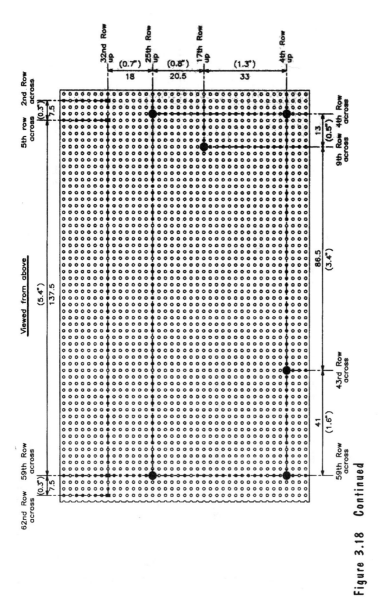

Figure 3.18 Continued

177

All dimensions in mm.

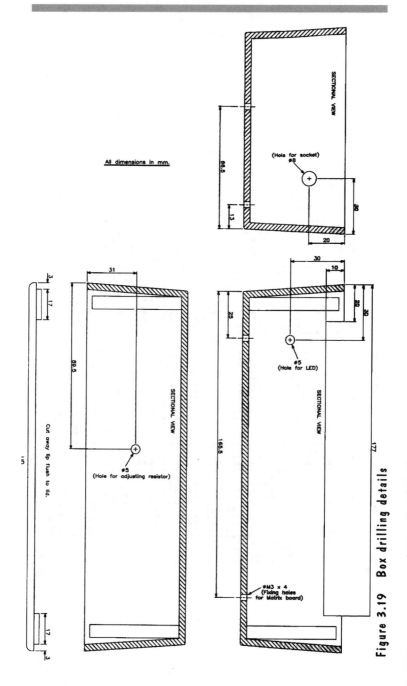

Figure 3.19   Box drilling details

At the front of the box a slot must be cut to allow bank-notes to be passed through. Part of the lip that runs around the box lid must also be removed. The inside lid of the box should be covered with white paper that fluoresces under UV light, (i.e. ordinary paper).

Refer again to Figure 3.18 for box assembly details. Four M3 x 30 mm countersunk bolts, M3 x $^1/_4$ in spacers; M3 nuts and M3 shakeproof washers are used to fix the module in place inside the box. Figure 3.20 shows wiring and final assembly.

Once installed, the LDR can be bent out over the top of the tube but with the sensitive face pointing towards the lid of the enclosure.

The LED should emerge from the front panel and the power socket from the right-hand side (when looking from the front). The buzzer can be bolted onto one of the fixing screws to stop it from moving around.

For increased reliability of the unit a sheet of transparent acrylic sheet may be fitted between the sensor and the bank-note (the PCB fixing screws may be lengthened by fitting 20 mm insulated spacers (FS38R) to allow it to be fixed). This will help to keep the bank-note flat when it is inserted, and enable the same distance to be maintained between the sensor and the bank-note each time it is used.

A front panel label is shown in Figure 3.21 and this can be photocopied or cut out and attached to the box.

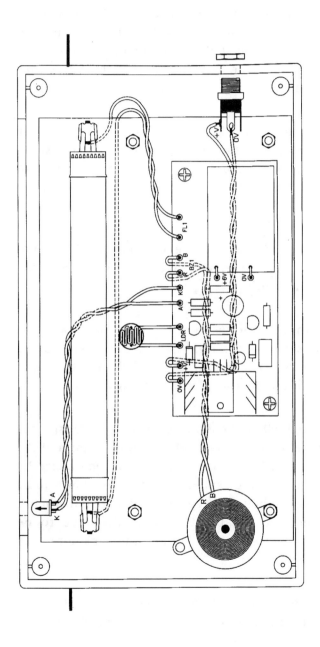

Figure 3.20 Final assembly and wiring

Calibration

Connect a suitable +12 V d.c. power supply to the unit, adjust the preset fully anti-clockwise, then insert a real bank-note through the aperture. If the unit fails to respond, adjust the preset clockwise. A point will be reached where the buzzer sounds and the LED will illuminate.

Test as many different value banknotes as possible; however, *old* banknotes may sometimes not pass the test. This is especially the case for ones that have accidentally been put through the washing machine. In other cases, some bank-notes have been *vandalised* with ultraviolet marker pens. However, a genuine note can be confirmed by inspection, as explained below.

Operation

It is quite easy to use the Till Saver. All you have to do is to connect a +12 V d.c. power source to the socket on the unit, this can be a +12 V d.c. car battery, or mains driven unregulated power supply, set for +12 V d.c. output. Once connected, insert the bank-note to be checked through the front of the unit. If the note is a real one, as it will be non-reflective to ultraviolet radiation, the LDR will detect a drop in reflected light level and the buzzer will sound and the LED will illuminate. If, however, the note is a fake, there will be little change in the light level, and so the buzzer will not sound and the LED will not illuminate.

Cut out circle A for LED

Figure 3.21   Front panel label

Important note

Please note that, although this project is designed to help detect counterfeit notes, Maplin Electronics plc cannot be held liable if the unit fails to do so. It should be used in conjunction with other methods of visually spotting counterfeit notes, such as making sure that there is a watermark, and that there is a silver thread interwoven through the note. The serial number should be in two places, and there should not be any blemishes in the printing.

Till Saver parts list

Resistors — All 0.6 W 1% metal film (unless specified)

R1	240 Ω	1	(M240R)
R2	910 Ω	1	(M910R)
R3	470 Ω	1	(470R)
R4	1k2	1	(M1K2)
R5	10 k	1	(M10K)
R6	100 Ω	1	(M100R)
R7	560 Ω	1	(M560R)
RV1	470 Ω vertical enclosed preset	1	(UH12N)

Capacitors

C1	100 nF miniature disc ceramic	1	(YR75S)
C2	1 µF 35 V tantalum	1	(WW60Q)
C3	220 µF 10 V miniature electrolytic	1	(JL06G)

Semiconductors

D1	1N4001	1	(QL73Q)
D2	1N4148	1	(QL80B)
LD1	LED green	1	(WL28F)
TR1,2	BC547	2	(QQ14Q)
RG1	LM317T voltage regulator	1	(UF27E)

Miscellaneous

LDR	ORP12 (light dependent resistor)	1	(HB10L)

Home security projects

	1 mm (0.4 in) single-ended		
	PCB pin	1	(FL24B)
	UV mini lantern	1	(ZC10L)
	heatsink	1	(FG55K)
BZ1	DC piezo buzzer	1	(CR34M)
	3 A 10 m wire red	1	(FA33L)
	3 A 10 m wire black	1	(FA26D)
	PCB	1	(GH76H)
	instruction leaflet	1	(XU61R)
	constructors' guide	1	(XH79L)

Optional (not in kit)

2.5 mm panel mount power		
socket	1	(JK10L)
M3 x 30 mm pozi screw	1	(JC72P)
M3 steel nut	1	(JD61R)
M3 shakeproof washer	1	(BF44X)
800 mA a.c. adapter		
unregulated	1	(YM85G)
type 3962 plain stripboard		
board	1	(JP53H)
11/4 in fuse clip	2	(KU28F)
M3 x 10 mm insulated spacer	1	(FS36P)
M3 x 1/4 in spacer	1	(FG33L)
type H2852 matt black ABS box	1	(BZ75S)
stick-on-feet	1	(FD75S)
2.5 mm standard power plug	1	(HH62S)
zip wire (bell wire)	as req	(XR39N)
large battery clip red	1	(FS86T)
large battery clip black	1	(FS87U)
5 mm LED clip	1	(YY40T)
quickstick pads	1	(HB22Y)

184